T0186840

Fluctuation Mechanism and Control on System Instantaneous Availability

Fluctuation Mechanism and Control on System Instantaneous Availability

YI YANG
Yong-Li Yu • Li-Chao Wang

CRC Press
Taylor & Francis Group
Boca Raton London New York

CRC Press is an imprint of the
Taylor & Francis Group, an **informa** business

CRC Press
Taylor & Francis Group
6000 Broken Sound Parkway NW, Suite 300
Boca Raton, FL 33487-2742

© 2016 by Taylor & Francis Group, LLC
CRC Press is an imprint of Taylor & Francis Group, an Informa business

No claim to original U.S. Government works

Printed on acid-free paper
Version Date: 20150911

International Standard Book Number-13: 978-1-4987-3467-7 (Hardback)

Library of Congress Cataloging-in-Publication Data

Yang, Yi, (Engineer) author.
   Fluctuation mechanism and control on system instantaneous availability / Yi Yang, Yong-Li Yu, and Li-Chao Wang.
      pages cm
   Includes bibliographical references and index.
   ISBN 978-1-4987-3467-7 (hardcover : alk. paper) 1. Systems availability--Mathematical models. 2. Reliability (Engineering)--Statistical methods. 3. Maintainability (Engineering)--Mathematical models. 4. Cycles--Mathematical models. 5. Control theory--Mathematics. I. Yu, Yong-Li, author. II. Wang, Li-Chao, author. III. Title.

TA169.Y364 2016
620.001'1--dc23   2015026121

Visit the Taylor & Francis Web site at
http://www.taylorandfrancis.com

and the CRC Press Web site at
http://www.crcpress.com

Printed and bound by CPI Group (UK) Ltd, Croydon, CR0 4YY

Contents

Preface

Equipment readiness and mission continuity are important elements of equipment operation effectiveness. Equipment availability is an important parameter describing equipment readiness and mission continuity. In actual use, equipment availability parameters are generally divided into steady-state availability parameters and instantaneous availability parameters.

Over the years, the research and applications of steady-state availability have been widely appreciated. However, with the increasing complexity of equipment functions, fault laws and protection laws of equipment are becoming more and more difficult to grasp. When new equipment systems come into use, there are often fluctuations of instantaneous availability due to interaction among subsystems; this is because the new equipment needs to break in at the initial service stage and cannot form battle effectiveness rapidly. The current research on equipment instantaneous availability fluctuation has not yet established a reasonable index evaluation system. Currently, research on equipment availability is basically around steady-state availability indicators, which reflect the relevant condition of equipment instantaneous availability when the time (t) approaches infinity. With the rapid development of technology, equipment updates and phase-out rates are accelerating; now, the service period of equipment may be only a few years or less, which is of more practical significance for research on equipment instantaneous availability during its service period. *Fluctuation Mechanism and Control on System Instantaneous Availability* tries to present a more complete theory and methodology as well as specific engineering implementation to lay the theoretical foundation for research on equipment instantaneous availability and to provide practical methods and tools.

With engineering needs as the background and the problems of the current research as the major breakthrough, this book starts from previous research work and focuses on the in-depth research on system instantaneous availability models as well as solutions, basic methodology, and framework for research on system instantaneous availability fluctuations within a certain range. The 10 chapters of this book are divided into basic and applied components. The basic chapters, Chapters 1–5, introduce the overview, the relevant concepts about system availability, the general continuous availability model approximate solution algorithms and problems, the discrete-time systems instantaneous availability model under general probability distribution, and the system instantaneous availability model and stability proof constraint by limited time, respectively. The remaining (applied) chapters focus on system instantaneous availability fluctuation analysis and control methods, discrete-time instantaneous availability model comparison and analysis under exponential distribution, system instantaneous availability fluctuations analysis under truncated discrete Weibull distribution, availability fluctuations optimal design under truncated discrete Weibull distribution, and end with a conclusion and outlook.

This book summarizes the latest research results and recent engineering practices at home and abroad. Many predecessors and counterparts in this field have worked hard to establish and develop this field. Without their work, this book could not have been written. To this end, we express our heartfelt thanks to our predecessors and their counterparts.

This book was compiled by Yi Yang, professor of aeronautics and astronautics, Beijing University, who also served as the chief editor, with Professor Yong-Li Yu and Senior Engineer Li-Chao Wang as coauthors. Professor Yi Yang was responsible for writing and compiling Chapters 1, 2, 3, 4, 6, 7, and 10 and for the compilation and modification of the whole book. Yong-Li Yu, an advocate of the system instantaneous availability research field, was responsible for proposing the book's framework and for assisting other authors with the completion of the relevant sections of the book. Li-Chao Wang, from the 28th Research Institute, was in charge of compiling Chapters 5, 8, and 9.

Special thanks to academician Sun Youxian. Despite having a busy schedule, he made many valuable and beneficial suggestions during the compilation of this book. His contribution was significant to the compilation. Special thanks also to Professors Zou Yun and Kang Rui, who provided in-depth and detailed guidance for our research work and offered many constructive comments during my doctoral thesis which were very helpful for the editors. Special thanks also to the National Natural Science Foundation for providing funding for the relevant research of this book.

Research on system instantaneous availability fluctuations is an emerging discipline with many unknowns, requiring practice and exploration on the part of the majority of the scientific and engineering staff. We hope that this book will play a valuable role, and we eagerly look forward to the reader's reaction.

Yi Yang
Beijing University

MATLAB® is a registered trademark of The MathWorks, Inc. For product information, please contact:

The MathWorks, Inc.
3 Apple Hill Drive
Natick, MA 01760-2098 USA
Tel: 508 647 7000
Fax: 508-647-7001
E-mail: info@mathworks.com
Web: www.mathworks.com

Authors

 **Yi Yang** earned a PhD at Nanjing University of Science and Technology, Nanjing, China, in 2008. She worked for five years at the Astronaut Research and Training Center, Beijing, China, and she was engaged in postdoctoral research at the School of Reliability and Systems Engineering, Beihang University (BUAA), Beijing, China, where currently she is an associate professor. She was a visiting scholar at the School of Computing and Mathematics, University of Western Sydney, Parramatta, Australia, for one year.

Her main research interests include reliability analysis and design, repairable systems, control science, and engineering. Her recent research work includes instantaneous availability modeling methods, fluctuations analysis, and control.

 **Yong-Li Yu** was born in 1964. He earned a BS at Shijiazhuang Mechanical Engineering College, Shijiazhuang, China, in 1982, and MS and PhD degrees at the College of East China Technology, Nanjing, China, in 1987 and 1990, respectively. Now he is a chair professor with the Maintenance Engineering Institute (MEI), Shijiazhuang, China, and an adjunct professor at Beihang University, Beijing, China. He has developed five courses and has published four books and more than 50 research papers. His main research interests include maintainability engineering and integrated logistics support.

 **Li-Chao Wang** is a senior engineer with the 28th Research Institute of China Electronics Technology Group Corporation, Nanjing, China. He earned a PhD at the School of Automation, Nanjing University of Science and Technology, Nanjing, China, in 2009, and worked as a postdoctoral researcher at the School of Reliability and Systems Engineering, Beihang University, Beijing, China.

His research interests are in the areas of system modeling, reliability analysis, performance evaluation, integrated logistics support, and quality control.

Chapter 1

Introduction

1.1 Research Background

Equipment operational readiness and continuity of mission importantly symbolize equipment combat power. As the important parameter measuring equipment operation readiness and continuity of mission, equipment availability is always given much attention by equipment developers and users. Generally, the equipment availability parameters are the steady state (such as equipment operational availability), achieved availability, and inherent availability.

Steady-state availability means that the whole equipment system gradually becomes steady after a long running-in period. Mathematically, this refers to the availability in case the operating time of the whole equipment system goes to infinity. Therefore, steady-state availability plays a very important role in our understanding of equipment operational readiness and continuity of mission through the whole service life of equipment. The steady-state availability parameters have been widely applied in various kinds of weapons to describe their operational readiness and continuity of mission, significantly promoting the equipment's construction. One of the most critical objectives in future combat system research is obtaining higher availability. Reference [1] studies how to improve the reliability and operating availability of a complicated system and proposes a pre-diagnosis-based method. A series of Monte Carlo simulations are proposed in Reference [1], concerning a complicated system consisting of 300 platforms, where every platform is comprised of 20 key units. In addition, the system availability curves under different strategies are also given.

The third generation of weapons and equipment that have been developed and used have functions and composition complexity as well as complexity in operation and maintenance far superior to those of the second generation. Because people know less about the fault mechanism and maintenance rules of third generation weapons and equipment, these weapons and equipment have very low availability in early deployment. For example, in the 1980s, only 27 of 72 U.S. Army F-15 fighters could fly in a readiness exercise with the U.S. early equipping force. The remaining fighters were forced to the ground due to insufficient spare parts, and their availability was only 37.5%. One F-15 needs to be maintained for 15 hours for every flight during the daily training, and only 9% of these F-15 fighters can fly continuously if they are repaired correctly in the workshop. Therefore, this F-15 is called the workshop queen so that it cannot become a combat power.

However, in the 1990s, F-15 availability reached 93.7% during the Gulf War. Obviously, this availability fluctuated heavily. Likewise, as our high-tech, complicated weapon systems have been increasingly developed over recent years, the same problem stands out as for the F-15 fighters. The availability of these complicated weapon systems usually fluctuates heavily in the early stage. But this fluctuation is quite far from the steady-state availability to be adopted for a long period by equipment developers and users. This factor seriously affects their understanding of the equipment availability level and the evaluation of the combat power of these complicated weapon systems. Conventional steady-state availability currently cannot be applied in the research on the new problem. A new kind of research must be developed to solve this urgent problem.

Instantaneous availability represents the availability level of the equipment at any operating point. It is a function of calendar time during equipment operation. Therefore, instantaneous availability can be used to describe and analyze the fluctuation that occurs during the equipment operation. The research on instantaneous availability fluctuation has two aspects—the mathematical probability model and field data statistics.

In field data statistics, a conclusion may be drawn using the definition of availability based on the basic theories and methodology of mathematical statistics (i.e., the instantaneous availability variation curve from a statistical rule view, such as the statistical values about the equipment availability of the F-15 in the early stages and during the Gulf War after a decade of use). This only means that the instantaneous availability fluctuates and cannot explain what causes the fluctuation and the occurrence principle using normal statistical research.

The mathematical model can generally answer the occurrence principle of the aforementioned fluctuation. Using the set-up model and found solution, the main principle of occurring fluctuation (or the main influence factors causing fluctuation) can be analyzed and studied under a specified parameter system. Therefore, for the purpose of guiding the engineering practice using theoretical research, this core model set-up is based on the principle conclusion that seeking fluctuation decrease can lay a solid foundation for further engineered research.

Apparently, this problem has not been studied using reliability mathematics, reliability engineering, and integrated equipment support. Therefore, it remains an important theoretical value. However, the pre-condition of applying the mathematical model to the research is that a mathematical model can be set up to describe this problem, proper analysis and solution methods can be found for this mathematical model, and it can extract and describe the parameter system of this fluctuation's characteristics. All of these conditions are the basis of research on equipment instantaneous availability fluctuation. This book focuses on setting up the model of instantaneous availability, especially in analyzing the solutions for this model. To help with this analysis, a brand new modeling method is outlined by selecting the typical instantaneous availability model to analyze the fluctuation occurrence principle, introducing the optimization of instantaneous availability during the equipment whole service life cycle, and determining the corresponding algorithm, thus laying the foundation for solving the aforementioned theories and engineering problems.

1.2 Research Status at Home and Abroad

Availability has been frequently studied using the basic theories of reliability mathematics, reliability engineering, and integrated equipment support [2–7] as is shown by the copious amount of literature on the subject. So far, the available research is mostly used for analyzing and estimating steady-state availability and average availability [8–11]. This research refers to the time proportions when the system is in normal state after the equipment runs for a long time. These two aspects have been well studied. Specifically, there are three main aspects.

The first is a study of how to set up an availability model and solve the steady-state availability with different system structures, unit life distributions, operation requirements, maintenance strategies, and kinds and quantities of maintenance resources; the second is to strictly mathematically prove that the steady-state availability exists under all research conditions, and third, to solely study the steady-state availability via Monte Carlo simulation.

As theories and engineering practices develop, instantaneous availability gradually interests more people in undertaking the research [12–20]. References [21] and [19] study the system availability model via the Markov process and renewal equation. References [17] and [22,23] study the instantaneous availability model for a certain special system via Laplace transform and inverse transform, respectively. References [14–16] use the concepts of failure rate and repair rate, defined within the global time domain, to set up the state transition model for system availability within the global time domain. Reference [18] systematically studies the approximate instantaneous availability model under incomplete repair conditions via simulation model and data analysis. References [24] and [13] systematically study the instantaneous availability models that may be ignored during part of the maintenance period. References [20] and [25,26] study the system availability models under the period monitoring strategy. Different from the unit service life, a virtual service life was proposed by Kijima, who successively built two kinds of models [27,28]. On this basis, References [18] and [29] study the system availability models under general maintenance or incomplete maintenance conditions.

The research in these references are model for the continuous-time system availability, using these instantaneous availability models to help predict and analyze whether there is steady-state availability and to accurately estimate the steady-state availability.

References [30–37] analyze the repairable system with unit failure time subjected to exponential distribution with the Markov process—the most common method for this kind of system availability analysis. When the time between failures and maintenance time of units is subjected to exponential distribution, these failures and maintenance of units are independent with each other; the Markov process can always be used to describe the analysis with properly defined system status. This repairable system is called the Markov repairable system [21]. The instantaneous availability model and its analysis have been labeled as more complicated, but the basic objective is still to try to solve and estimate the steady-state availability of quite complicated systems.

Beside the Markov process, the random Petri-net model is also often used in steady-state availability analysis for repairable systems due to its good mathematical nature. References [38,39] propose a simple Petri-net-based method to calculate the steady-state availability for repairable redundant systems.

However, the models in these references all have a pre-condition: the repairing time and the maintenance time are subjected to exponential distribution. As we know, actually there are a lot of units with a failure time or maintenance time that does not follow the exponential distribution. For example, the failure time of relays and switches is subjected to the Weibull distribution, while their maintenance time is subjected to log-normal distribution; the failure time of transformers is subjected to normal distribution as well. Therefore, these models are somewhat limited.

Meanwhile, the Markov process generally has a state space explosion in cases of complicated repairable systems. In light of this, Reference [10] describes the dynamic time sequence of the system via dynamic fault tree logic and transforms it to the corresponding Markov state transition chain to calculate the equivalent failure rate and repair rate of each independent sub-tree and to recourse them into the overall fault tree. In this way, it can avoid the complicated and possible problems occurring during set-up of the Markov chain graph and can reduce the quantity of solved differential equations, thus decreasing the calculation load and analysis difficulty and accelerating the calculation speed.

For those complicated repairable systems where the failure time or the repair time is not subjected to the exponential distribution, their availability is generally analyzed with simulation techniques [40–41]. Among these simulation techniques, the Monte Carlo is the most commonly used simulation technique. Reference [42] gives an example of a computer simulation method for calculating the availability of a double-unit cold standby radar system where the repair time is subjected to the normal distribution.

Reference [12] designs a Monte Carlo method for evaluating the reliability and instantaneous availability for complicated repairable systems. It uses the bootstrap as the framework to evaluate the reliability and instantaneous availability of a non–Monte Carlo system with period maintenance. Reference [43] describes the operating state of a cluster system and then quantitatively processes the structural function relations among nodes, giving the probability distribution and sampling method of random variables on each node in the cluster system. Then, based on the identification results from a failure occurrence state and a repair state, the operating state of each node will be confirmed, so as to establish a logical relationship for availability simulation and to obtain the availability estimate of the cluster system. References [44,45] analyze the operational availability of aircraft in a simulation. The unit failure can be caused by its own failure and battle damage, and these units have delay-repair time. In the models from References [46,47], the units of subsystems have their failure time subjected to Weibull distribution, their maintenance time subjected to exponential distribution, and their delay-repair time subjected to normal distribution.

The purposes of the aforementioned research are almost the same as other research literature about instantaneous availability, that is, better calculating and estimating the steady-state availability of complicated systems. In addition, the probability distributions in this research do not really conform to the realities of complicated systems.

Many scholars have studied the availability equation stability via functional analysis and other analytical tools. Reference [48] proves the existence, uniqueness and nonnegativity of solutions to two different-unit repairable systems with a strongly continuous operator semigroup, and it obtains this system stability by studying the spectral features of corresponding operators. References [49,50] study a two-unit parallel system and a two-different-unit series system, respectively. Considering that the system solutions exist and are nonnegative and unique, it is proven by the compressed C_0 semigroup in a Banach space generated by system operators that the nonnegative stable solution of the system is just the nonnegative eigenvector corresponding to the 0 eigenvalue of the system operator. Also, by studying the spectral features of system operators, it has been proven that the spectral points of system operators are all located on the imaginary axis on the left half of a complex plane (except for 0 with no spectrum), thus obtaining the system asymptotic stability.

Reference [51] proves the existence and uniqueness of nonnegative solutions to two-unit series repairable systems in a purely analytical way, and it also studies the monotone stability of systems. Reference [52] proves the existence and uniqueness of nonnegative solutions to two same-unit cold standby repairable systems with a strongly continuous operator semigroup. Reference [53] proves the existence and uniqueness of nonnegative solutions to two same-unit warm standby repairable man-machine systems with a strongly continuous operator semigroup; 0 is the eigenvalue of system main operators, and it has the corresponding normal eigen-vector, which proves the system has steady-state normal solution.

Reference [54] discusses the mathematical models for the two same-unit parallel repairable systems: estimate the resolvent of the main operator, prove the stability of the main operator, prove the existent uniqueness of nonnegative solutions to paral-lel repairable systems with strongly continuous operator semigroup, and describe how the system operators generate a strong, continuous, and normal operator semi-group. Reference [55] discusses the mathematical models for the dynamic series repairable systems, proving the existence and uniqueness of nonnegative solutions to this system through C_0 semigroup in the functional analysis and further con-firming the solution is semistable under certain conditions. Reference [56] proves the existence, uniqueness, and monotone stability of dynamic nonnegative solution to the single-unit repairable system in a purely analytical way.

Reference [57] analyzes the spectral features of system main operators via functional analysis and proves the asymptotic stability of solution to the repair-able man-machine system with spare parts. Reference [58] proves the existence and uniqueness of nonnegative solution to the two same-unit parallel repairable sys-tems with elementary methods. Reference [59] discusses the mathematical models for the two same-unit parallel repairable system, applying the strongly continuous operator semigroup to prove the existence and uniqueness of nonnegative solu-tion to the system and to confirm the system's stability. Reference [60] proves the existence and uniqueness of dynamic nonnegative solution to the four-unit (having common-cause failures) redundant repairable system in a purely analytical way.

Based on the strongly continuous operator semigroup theory, Reference [61] proves that the system consisting of the robot and attached security guard has the unique nonnegative dynamic dependent solution. In addition, the system has the normal steady-state solution under certain conditions, and its dynamic solution is typically (in space norm context) asymptotically converged to the steady-state solution. Reference [62] considers the series repairable systems. When their units malfunction and cannot be repaired in time for some reason, a delay-repair time period occurs. Then the whole renewal process is as follows: normal operation (until failures occur), maintenance delay, normal operation, and so forth. It is assumed that every period distribution is unknown and that they are independent of each other; the availability point estimate and confi-dence lower limit are given.

Despite the fact that the aforementioned research addresses the variation trend of instantaneous availability, the research is limited to specifically mathematical analysis, thoughts, and methods. All analyze and estimate the asymptotic behavior of instantaneous availability when the time (t) goes to infinity but do not address the fluctuation and variation rule of instantaneous availability during a limited period at the preliminary operation of the system. The latter is the focus of this book.

Research on the availability have made very impressive advances over the years. However, there is still a lot to be studied with respect to the demands of instantaneous availability proposed during the actual equipment operation and maintenance. To be specific, there are four aspects.

1. The modeling methods for instantaneous availability have been well studied; scholars have set up complete instantaneous availability equations (differential, partial differential, or integral equations), but these models are mostly used to study whether there is steady-state availability or to solve the steady-state availability. Most research on solving the instantaneous availability only gives the exponential time distribution and does not solve the instantaneous availability under the general probability distribution. In other words, the current research has not really systematically studied the solution of instantaneous availability equations. Therefore, the fluctuation characteristics seen from the instantaneous availability equations have not been addressed because the research range is only limited to the existence of steady-state availability in cases where the time goes toward infinity and the calculation of steady-state availability.

2. By applying the functional analysis and pure analysis methods strictly mathematically, the instantaneous availability of multiple repairable systems is proven. When the time goes toward infinity, the instantaneous availability is stable or asymptotically stable. These conclusions have laid a firm theoretical foundation for the research and application of steady-state availability. However, the actual engineering concerns influence the fluctuation characteristics represented by the instantaneous availability during the preliminary operation of the equipment. The methods applied by currently established instantaneous availability equations to analyze this fluctuation problem really are more concerned with engineering. Strictly speaking, there is no real significance in having the time go toward infinity in the engineering practices because the equipment (or the product) cannot be infinitely used. Therefore, the important basic theories regarding the stability of instantaneous availability at certain time intervals is what urgently needs to be studied.

3. The current research on instantaneous availability is all about continuous-time systems, and the set-up instantaneous availability models are all continuous-time models. Basically, there are no problems related to discrete-time systems or to discrete-time models. In fact, in many engineering practices, the occurrence and termination of failures show the characteristics of discrete-time system, as well as the maintenance at the beginning and the end. The frequent inspection of system period, preventive maintenance/time for replacing units, and the repair time may not always be processed as continuous random variables. They can be considered as the sampling sequences of random variables in nonnegative integer values as properly assumed on the real conditions. In this case, a new discrete-time instantaneous availability model system can be set up for the instantaneous availability research.

Thus, scholars can move from the continuous-time model research to the research on the stability of instantaneous availability during certain intervals.

4. Focusing on how to control the variation rule of instantaneous availability on certain intervals is the main starting point for us in studying the fluctuation of instantaneous availability. For these systems, there may be many kinds of different fluctuation rules of instantaneous availability that correspond with one steady-state availability value. In terms of engineering demands, the apparent goal is to obtain the same steady-state availability level with less fluctuation.

1.3 Main Content and Structure

In order to meet the aforementioned actual demands of instantaneous availability and analyze its existing problems, this book focuses on the engineering demand, preliminarily developing the basic research methods for instantaneous availability fluctuation on certain intervals on the basis of previous research results. Therefore, the first problem is to solve the instantaneous availability equations. Second is to analyze the fluctuation characteristics of instantaneous availability on certain intervals after solving its equations and to properly analyze and design a way to control this fluctuation.

It should be noted that the instantaneous availability fluctuation on certain intervals is a new research field with many problems to be addressed. Because this book is a preliminary discussion, not all the problems can be solved here. Therefore, the focus of this book is on the solution analysis of conventional instantaneous availability equations, new modeling methods for instantaneous availability, simulation analysis of fluctuation principle of instantaneous availability on certain intervals, and on the optimal control over the instantaneous availability fluctuation on certain intervals.

In addition, it should also be pointed out that this is not just a single research point but an attempt to provide connectivity for this whole field of research. The preliminary research on some core parts will provide a referable research framework for the future. Therefore, the main research content covers the following:

1. Deep and extensive research based on the existing instantaneous availability models of a continuous-time repairable system to set up a typical instantaneous availability equation under general probability distribution of which the maintenance time and logistic delay time in system failure maintenance, corrective maintenance, and preventive maintenance need to follow the general probability distribution, and to solve this complicated equation by using the successive approximation method, thus obtaining the approximate solution of system instantaneous availability of the approximate availability of the system. In addition, this book will analyze the physical meaning

of the approximate solution of system instantaneous availability, discuss the possibility of replacing the accurate solution of instantaneous availability with its approximate availability, and then will give corresponding conclusions and suggestions.

2. Analysis and discussion of the existing problems in continuous-time instantaneous availability models and the discrete-time characteristics of system real maintenance will be covered as well as considering that the repair time and logistic delay time in system failure maintenance, corrective maintenance, and preventive maintenance need to follow the general discrete probability distribution and set up the state transition models for general discrete-time repairable systems, discrete-time repairable systems with repair delay, and discrete-time repairable systems with preventive maintenance. Using numerical examples, we will study the newly built models and then conduct comparative analysis between these models and the continuous-time models having the same problems as well as sort out their advantages, disadvantages, and interrelations, and discuss the equivalence between these two kinds of models on the same system. Meanwhile, we will prove the stability of system instantaneous availability using matrix theory and methods, namely the existence of system steady-state availability, and then obtain the expression of steady-state availability.

3. We will propose parameters describing the fluctuation characteristics of system instantaneous availability within a limited time, which well reflects the fluctuation level of system instantaneous availability to some extent. Using this basis, the optimal control model with the fluctuation parameters will be set up.

4. The MATLAB® numerical calculation tool will be used to conduct numerical simulation on the typical discrete-time instantaneous availability model. We will then propose the system optimization for instantaneous availability and constraining steady-state availability and describe different expressions for this optimization during the whole service life of the equipment. Typical instantaneous availability models will be selected and the typical optimization problem will be solved using the particle swarm optimization algorithm to determine suitable methods in relevant engineering practices.

5. This book will analyze the instantaneous availability of repairable systems via simulation under Weibull truncated distribution. Considering the case that steady-state indexes are fixed (such as mean time to repair, mean time between failures, or mean logistic delay time), the relationship between the system fluctuation parameters and parameters related to time distribution will be studied. Finally, the variation laws of fluctuation parameters corresponding to relevant variables will be obtained.

6. Under Weibull truncated distribution conditions, the optimal control models designed with availability fluctuation characteristic are degraded to the constrained multi-variable optimization models, including the minimum

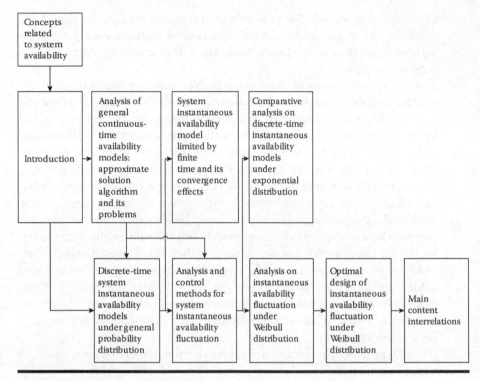

Figure 1.1   Main content interrelations.

availability amplitude model, system optimal matching model, and optimal preventive maintenance cycle model. These optimal models are studied in the demonstration, development, and operation phases, respectively, during the whole service life of the equipment. Also, simulation analysis is conducted to validate the effectiveness of particle swarm optimization for the models. Based upon the main research content, this book consists of 10 parts, as shown in Figure 1.1.

Chapter 2

Concepts Related to System Availability

Research on instantaneous availability involves some basic theories and methods with respect to reliability engineering, reliability mathematics, random process, particle swarm algorithms, and so forth. In order to more easily understand the research on instantaneous availability, this chapter presents a brief introduction to the basic concept of availability; definition of main availability parameters; basic concepts of reliability, maintenance, and supportability, definitions of their main parameters, basic models for availability, and so on.

2.1 Basic Knowledge about Reliability

2.1.1 Basic Concept of Reliability

The term reliability means the capability of a product to complete its specified functions under certain conditions within a finite period of time. The metric unit for probability is reliability. The opposite of reliability is product failure or invalidation. Strictly speaking, specified function loss is caused by invalidation or failure. Generally, the unrepairable products are called fault, and the repairable products are called failure. The product failure time (or service life) is affected by many factors, such as the materials used for this product, conditions occurring during design and manufacturing processes, and the environmental conditions for storing and using the product. The failure time or the service life of the product may also be associated with the functions it needs to complete. When the product loses its specified functions, it becomes invalid, so its service life comes to an end.

Obviously, the service life of one particular product used under the same environmental conditions may vary due to different specified functions.

Suppose $t \geq 0$ is the specified time, and random variable $X \geq 0$ is the operating time before the product failure. The distribution function

$$F(t) = P\{X \leq t\} \qquad (2.1)$$

represents the probability of losing specified functions within time $[0, t]$, which is called the accumulated failure probability or unreliability of the product. Function

$$R(t) = \bar{F}(t) = P\{X > t\} \qquad (2.2)$$

represents the probability of completing the specified conditions within time $[0, t]$, which is called the reliability function or the reliability of the product.

Conditional probability of product failure per unit time after it operates normal till time (t),

$$\lambda(t) = \lim_{\Delta t \to \infty} \frac{P\{t < X \leq t + \Delta t \mid X > t\}}{\Delta t}. \qquad (2.3)$$

Equation 2.3 is called the failure rate function of a product, describing the variation rule of product failure following the up-time change.

The interrelations among the failure rate function $\lambda(t)$, reliability function $R(t)$, accumulated failure distribution function $F(t)$, and failure distribution density function $f(t)$ are shown as follows:

$$\lambda(t) = \frac{F'(t)}{R(t)} = \frac{f(t)}{R(t)} = -\frac{R'(t)}{R(t)} \qquad (2.4)$$

$$R(t) = e^{-\int_0^t \lambda(t)dt}. \qquad (2.5)$$

2.1.2 Main Parameters Describing Reliability

For unrepairable products, the main quantitative parameters of reliability are reliability $R(t)$, failure rate $\lambda(t)$, and mean service life $(MTTF)$. Suppose the product operates normally from $t = 0$. If X is its service life, its operating process as time elapses will be shown as in Figure 2.1. Because the product has no reason to be repaired, it will remain permanently in the invalidation state once it becomes invalid. Meanwhile, the definition formulas of reliability and its mean service life also describe the reliability characteristics of the product.

For the repairable products, it is more complicated. Because the product is repairable, it can be repaired after its failure. At this time, the product operates normally or,

Figure 2.1 **Unrepairable product.**

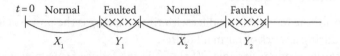

Figure 2.2 **Repairable product.**

alternatively, with failures as the time elapses. As shown in Figure 2.2, X_i and Y_i represent the product up-time and down-time in ith cycle, respectively, where i = 1, 2, ... The product operates normally within up-time and is in a failure state within down-time. Generally, X_1, X_2, ... or Y_1, Y_2, ... may not be under the same distribution. In addition to using the failure rate function and reliability function to describe the quantitative indexes of reliability for repairable products, there are mean time to the first failure, mean failure frequency, and mean time between failures.

2.1.2.1 Mean Time to Failure

The failure time of a given product under specified conditions with specified functions and within specified time [0, t], using the random variable $X \geq 0$, is

$$EX = \int_0^\infty t dF(t), \qquad (2.6)$$

which is the mean time to failure of the product ($MTTF$) or mean time between failures ($MTBF$).

2.1.2.2 Mean Time to the First Failure

The distribution of the mean time to the first failure X_1 for the product is defined as

$$F_1(t) = P\{X_1 \leq t\}. \qquad (2.7)$$

The mean time to the first failure ($MTTFF$) is defined as

$$MTTFF = EX_1 = \int_0^\infty t dF_1(t). \qquad (2.8)$$

This represents the probability that the repairable product operates normally during [0, t], consistent with the general definition of the aforementioned reliability.

Once a repairable product fails, which will cause disastrous consequences, both the distribution and mean value of the *MTTFF* are the important quantitative indexes for the reliability.

2.1.2.3 Mean Failure Frequency and Steady-State Failure Frequency

The process for the repairable product with time elapsing is the process alternatively occurring during a series of normal failures.

Suppose the random variable $N(t) \in \mathbb{Z}$ is the failure frequency within $[0, t]$. The distribution is as follows:

$$P_k(t) = P\{N(t) = k\}, \quad k = 0, 1, 2, \ldots \tag{2.9}$$

The mean failure frequency for the product within $[0, t]$ is

$$M(t) = EN(t) = \sum_{k=1}^{\infty} k P_k(t). \tag{2.10}$$

When $M(t)$ derivative exists,

$$m(t) = \frac{d}{dt} M(t) \tag{2.11}$$

is the instantaneous failure frequency of the product.

2.1.2.4 Mean Time between Failures

Suppose failures of the repairable product occur n times during its operation. After every failure has been repaired, this product operates continuously. Its up-times are $t_1, t_2, \ldots t_n$; then the *MTBF* is defined as

$$MTBF = \frac{1}{n} \sum_{i=1}^{n} t_i. \tag{2.12}$$

2.2 Basic Knowledge about Maintenance and Supportability

2.2.1 Concept and Main Parameters of Maintenance

Maintenance refers to maintaining or recovering the competence of a product while repairing it in accordance with regulated procedures and methods under specified conditions and within a set time. Major maintenance parameters involved in instantaneous availability are as follows.

Suppose random variable $Y \geq 0$ is the repair time of a product. The repair time distribution function is

$$M(t) = P\{Y \leq t\}, \ t \geq 0. \tag{2.13}$$

This is called product maintainability.

Maintainability can reflect the probability of Y (the actual repair time) less than or equal to t (the regulated repair time). Obviously, $M(t)$ is a probability distribution function.

The probability density function of product repair time refers to the ratio of repaired quantity and total quantity delivered to be repaired within unit time, which is the probability of products expected to be repaired per unit time:

$$m(t) = \frac{dM(t)}{dt} = \lim_{\Delta t \to 0} \frac{M(t + \Delta t) - M(t)}{\Delta t}. \tag{2.14}$$

2.2.1.1 Product Repair Rate

$$\mu(t) = \lim_{\Delta t \to \infty} \frac{P\{t < Y \leq t + \Delta t \mid Y > t\}}{\Delta t}. \tag{2.15}$$

Product repair rate can reflect the probability of products that cannot be repaired at t but that can be repaired within unit time after that time. Obviously,

$$\mu(t) = \frac{m(t)}{1 - M(t)}. \tag{2.16}$$

2.2.1.2 Mean Time to Repair

Suppose the random variable $Y \geq 0$ is the repair time for a specified product. Mean time to repair ($MTTR$) for the product is

$$EY = \int_0^\infty t dM(t). \tag{2.17}$$

In the product maintenance process, maintenance types are generally divided into repairable maintenance and preventative maintenance. The aforementioned maintainability parameters are also applicable to these two types.

2.2.2 Concept and Main Parameters of Supportability

Supportability refers to the ability of the system (i.e., the equipment) design characteristics and planned support resource to meet the operating requirements for daily use and battle use. We can see from the definition that the scope of supportability ranges from the equipment itself to the equipment support system. From the perspective of instantaneous availability study, involved parameters are mainly a description of support-system operation.

2.2.2.1 Probability Distribution Function of Logistic Delay

Suppose random variable Z $(Z \geq 0)$ is the product logistic delay time. The distribution function is

$$S(t) = P\{Z \leq t\}, t \geq 0. \tag{2.18}$$

This is called the probability distribution function of accumulated logistic delay.

The probability distribution function of accumulated logistic delay can reflect the probability of Z (the duration of completing logistic delay) being less than or equal to t (the regulated logistic delay time under certain conditions).

2.2.2.2 Probability Density Function of Logistic Delay Time

$$s(t) = \frac{dS(t)}{dt} = \lim_{\Delta t \to 0} \frac{S(t + \Delta t) - S(t)}{\Delta t} \tag{2.19}$$

2.2.2.3 Switch-to-Repair Rate

$$\rho(t) = \lim_{\Delta t \to 0} \frac{P\{t < Z \leq t + \Delta t \mid Z > t\}}{\Delta t} \tag{2.20}$$

Switch-to-repair rate can reflect the probability of product at t being switched to repair state within unit time after t.

2.2.2.4 Mean Logistic Delay Time

Suppose random variable $Z \geq 0$ is the mean logistic delay time for a specified product; the mean logistic delay time ($MLDT$) is as follows:

$$EZ = \int_0^\infty t\, dS(t). \tag{2.21}$$

There are many other parameters that can reflect the operation of support systems besides the parameters or functions reflecting the maintainability and supportability of repairable products.

2.3 Frequently Used Probability Distributions

2.3.1 Geometric Distribution

Suppose the random variable X follows the geometric distribution, if

$$p_k = \Pr\{X = k\} = pq^k, \quad k = 0, 1, 2, \ldots \tag{2.22}$$

where $p, q > 0$, $p + q = 1$. If X is the system service life, then the failure rate is

$$\lambda(k) = p, \quad k = 0, 1, 2, \ldots \tag{2.23}$$

In fact, geometric distribution is the only distribution containing a "lack-of-memory property" in discrete distribution. It is similar to continuous exponential distribution.

2.3.2 Discrete Weibull Distribution

Suppose nonnegative integer random variable X follows

$$p_k = \Pr\{X = k\} = q^{k^\beta} - q^{(k+1)^\beta}, \quad k = 0, 1, 2, \ldots \tag{2.24}$$

$0 < q < 1, \beta > 0$

If X follows the discrete Weibull distribution whose scale parameter is q and shape parameter is β and X refers to system service life, its failure rare is as follows:

$$\lambda(k) = 1 - p^{(k+1)^\beta - k^\beta}, \quad k = 0, 1, 2, \ldots \tag{2.25}$$

When $\beta \geq 1$, $\lambda(k)$ monotonically increases; when $\beta \leq 1$, $\lambda(k)$ monotonically decreases. When $\beta = 1$ and $p_k = q^k - q^{k+1} = (1 - q)q^k$, the Weibull distribution is degraded to geometric distribution.

2.4 Concept of Availability

2.4.1 Basic Concept of Availability

Availability is a parameter that turns the comprehensive measurement of system reliability, maintainability, and supportability into efficiency. It represents whether the product is available when needed under regulated conditions, which also refers to the combat power at any time. The probability measurement of usability becomes availability. This parameter is an integrated index considering maintainability, reliability, and supportability as a whole.

2.4.2 Major Parameters of Availability

2.4.2.1 Instantaneous Availability

Suppose a repairable system $X(t)$ has two possible states: normal and failure. When $t \geq 0$,

$$X(t) = \begin{cases} 0 & \textit{System normal at the time of } t \\ 1 & \textit{System failed at the time of } t \end{cases}$$

$$A(t) = P\{X(t) = 1\}. \tag{2.26}$$

Instantaneous availability $A(t)$ only means whether the product is normal at t or not. It does not include whether the product had a failure before t or not.

The mean availability within $[0, t]$ can be defined as

$$\tilde{A}(t) = \frac{1}{t} \int_0^t A(u)du, \tag{2.27}$$

where $A(t)$ is the instantaneous availability defined in (2.26). If the limit $A = \lim\limits_{t \to \infty} A(t)$ exists, regard it as steady-state availability; consider $\tilde{A} = \lim\limits_{t \to \infty} \tilde{A}(t)$ as extreme mean availability; and $\tilde{A} = A$.

Steady-state availability is generally divided into inherent availability (A_i), achieved availability (A_a), and operational availability (A_0). These three concepts are the steady-state availabilities that are important in different stages of engineering during the whole lifetime of the equipment. Computing methods for these three steady-state availabilities are as follows.

2.4.2.2 Inherent Availability (A_i)

Under certain conditions, system availability is required to be set by working hours and repairable maintenance. Availability regulated by this means is called inherent availability:

$$A_i = \frac{T_{BF}}{T_{BF} + \bar{M}_{ct}}, \tag{2.28}$$

where T_{BF} is mean time between failures, while $\bar{M}_{ct}$ is mean time to repair.

2.4.2.3 Achieved Availability (A_a)

The measurement of achieved availability A_a can be calculated by (2.29):

$$A_a = \frac{T_0}{T_0 + T_{CM} + T_{PM}}, \tag{2.29}$$

where T_0 is working time, T_{CM} is repairable maintenance time, and T_{PM} is preventative maintenance time. A_a is also determined by the product structural performance; and preventative maintenance also will be considered.

2.4.2.4 Operational Availability (A_0)

When *MTBF* and mean down-time (MDT) are known, operation availability A_0 is

$$A_0 = \frac{T_{BM}}{T_{BM} + T_D}, \tag{2.30}$$

where T_{BM} is mean time between failures and T_D is mean down-time.

In general, operational availability A_0 is impacted by product utilization rate. Within a set time, the shorter the system working time, the higher A_0 becomes. Therefore, the down-time or comparatively short working time (e.g., repair and storage time) of the system shall not be included when determining the total time of the system.

Formula 2.30 is applicable to developing initial parameter determination and sensitivity analysis. When using this formula, suppose the standby time (mean set-up time) is zero.

In addition, when *MTBF*, *MTTR*, and *MLDT* are obtained, A_0 can also be calculated as follows:

$$A_0 = \frac{T_{BF}}{T_{BF} + \bar{M}_{ct} + T_{MLD}}, \tag{2.31}$$

where T_{MLD} is mean logistic delay time (including the time spent in waiting for spare parts, maintenance personnel, and support equipment).

2.5 Common Problems with System Instantaneous Availability

Common problems with system instantaneous availability can be divided into those of single-unit repairable systems and those of multi-unit repairable systems; the problems of multi-unit repairable systems can be further divided into those of multi-unit series repairable systems, multi-unit parallel systems, multi-unit voting repairable systems, and of multi-unit cold standby repairable systems, according to common system availability models.

2.5.1 Single-Unit Repairable System

A single-unit repairable system is the simplest repairable system. A system generally consists of one unit and one repair worker. The system works when the unit works, and the system breaks down when the unit breaks down. The broken-down unit enters a working state right after it is repaired, and the system enters a working state correspondingly. So, there are two states for the system: system failure and system working state. If the unit can be repaired perfectly, with X representing the unit breakdown time and Y representing the repair time after the unit breaks down—two random variables independent from each other, their distribution functions and expectations are as follows:

$$F_X(t) = P\{X \le t\}, \qquad E[X] = \int_0^\infty t \, dF_X(t) = \frac{1}{\lambda} \tag{2.32}$$

$$G_Y(t) = P\{Y \le t\}, \qquad E[Y] = \int_0^\infty t \, dG_Y(t) = \frac{1}{\mu}, \tag{2.33}$$

where $\lambda > 0$, $\mu > 0$. The problem with system instantaneous availability study is establishing an update process alternately appearing in a system working state and a repair state by a theory of random processes in accordance with the situation occurring alternatively in system states. And then the problem is applying the basic theories of the update process and the related parameter definition mentioned earlier to determine the system instantaneous availability equation, mean time to first failure, mean system failure frequency within $(0, t)$, and system steady-state availability.

Considering the logistic delay of a single unit, we must introduce the distribution function and expectation of logistic delay time W:

$$H_W(t) = P\{W \le t\}, \qquad E[W] = \int_0^\infty t \, dH_W(t) = \frac{1}{\rho}. \tag{2.34}$$

There are three states alternatively appearing in the system at this moment: system working state, failure waiting state, and system maintenance state. The update process of these three states according to theory of random processes must be established for the study of system instantaneous availability problem.

2.5.2 Multi-Unit Repairable System

The failure repair and logistic delay of a multi-unit repairable system are similar to those of a single-unit repairable system. The main difference lies in the descriptions of system failure and a single unit.

Series system refers to a system consisting of n different units and one repair worker. If the n units can work normally, the system will be in a working state. When one unit breaks down, the system will be in a failure state. The repair worker shall immediately repair the broken-down unit, and all other units shall be stopped.

When the broken-down unit is repaired, all units enter the working state at once and the system does the same accordingly.

Parallel system refers to a system consisting of n different units and one repair worker. The system can work when all those n units can work or at least two units can work. The system only enters failure sate when all the n units break down. Once one unit breaks down, the repair worker shall immediately repair it. The broken-down unit enters a working state right after it is repaired. If one broken-down unit is not repaired and another one breaks down, then the second one enters a to-be-repaired state.

Suppose all the n units can be repaired as new after each failure, while $X_i (i = 1, 2, \ldots, n)$—the failure time for the ith unit and $Y_i (i = 1, 2, \ldots, n)$—the repair time for the broken-down ith unit is represented by random variables independent from each other; the distribution functions and expectations for these two variables are as follows:

$$F_{X_i}(t) = P\{X_i \leq t\}, \qquad E[X_i] = \int_0^\infty t dF_{X_i}(t) = \frac{1}{\lambda_i} \tag{2.35}$$

$$G_{Y_i}(t) = P\{Y_i \leq t\}, \qquad E[Y_i] = \int_0^\infty t dG_{Y_i}(t) = \frac{1}{\mu_i}. \tag{2.36}$$

The study of multi-unit repairable system instantaneous availability analyzes the system state and the situation arising from the system state; establishes the update process alternatively appearing between system working state and system repair state by theory of random processes; and determines system instantaneous availability equation, mean time to first failure, mean system failure frequency within $(0, t)$, and system steady-state availability.

Note: The repair worker may be busy or the to-be-repaired phenomenon may appear in the repairable systems with redundancy as mentioned earlier. However, those phenomena are not same as logistic delay. Therefore, logistic delay shall also be considered in a single-unit repairable system.

2.6 Summary

This chapter briefly discusses some related theories, technical methods, and ideas that are the precondition and basis of the study.

Chapter 3

Approximate Solution and Its Problem for General Continuous Availability Model

In order to study the law of fluctuating variation of instantaneous availability, we first must establish the instantaneous availability equation and solve it. For our purposes, it is better to obtain an analytical solution to the equation. This problem has been a hot topic in the mathematics research field for a long time. Generally speaking, when the fault time and repair time of each system unit and other related distributions are all exponential distributions, the system can be described using the Markov process only if the status of system is properly defined [63–65]. In References [8] and [66–69], the integrated analytical solution of instantaneous availability of a Markov repairable system is obtained. When at least one related distribution in a repairable system is not of exponential distribution, the system status can be described with renewal process theory and the instantaneous availability of a system meets one renewable equation. Many scholars have determined the instantaneous availability equation of a repairable system with general probability distribution for theories related to application of the renewal process have also studied equation reliability, which ensures the steady-state availability of a repairable system but does not obtain the analytical solution of instantaneous availability. References [17,70] show that the instantaneous availability of a system with service life of gamma distribution and repair time of exponential distribution is related to Laplace transform. However, the analytic solution of instantaneous availability of a repairable system with general probability distribution has not been given in any reference, which causes difficulty in the research on fluctuation of instantaneous availability.

Because it is difficult to obtain the analytic solution of system instantaneous availability with general probability distribution, we can try to find out a proper method to obtain an approximate solution, which is also important for solving the aforementioned problem. In this chapter, the primary focus is as follows: Under the condition of general probability distribution and single-unit system, seeking the approximate solution of an instantaneous availability equation, analyzing its error, and demonstrating whether the approximate solution is applicable for engineering.

3.1 Availability Equation Analysis of Single-Unit System with General Probability Distribution

Suppose the system is composed of one single unit and its fault time X conforms to general probability distribution $F(t)$. The unit is repaired immediately after it is out of order, and the repair time Y also conforms to general probability distribution $G(t)$. After having been repaired, the unit immediately returns to normal working status. Therefore, all expectations are as defined in (2.32) and (2.33). After repair, the service life of the unit is the same as that of a new one; X and Y are mutually independent.

For the sake of simplicity, assume the unit is a new one at the initial stage (i.e., the unit enters the working status when $t = 0$). Consider one renewal process (alternately working and being repaired), as shown in Figure 3.1. If we make $Z_i = X_i + Y_i$, X_i and Y_i refer to the service life and repair time, respectively, of unit in cycle i if $i = 1, 2, \ldots, \{Z_i, i = 1, 2, \ldots\}$ is a string of independently and identically distributed random variables in sequence. They compose a renewal process. The renewal life is as follows:

$$Q(t) = P\{Z_i \leq t\} = P\{X_i + Y_i \leq t\}$$

$$= \int_0^t G(t-u)dF(u) = F(t)*G(t), \quad i = 1, 2, \ldots \qquad (3.1)$$

From this figure, it can be seen that the time is X_1 before the first failure of system. So, the system reliability is

$$R(t) = P\{X_1 > t\} = 1 - F(t). \qquad (3.2)$$

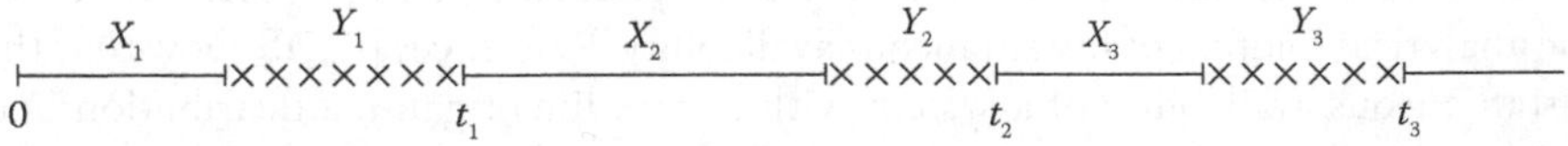

Figure 3.1   Renewal process with alternate appearance of working and repairing.

In order to determine the instantaneous availability of a system, we introduce a random process $\{X(t), t \geq 0\}$, in which

$$X(t) = \begin{cases} 1, & \textit{System is working at the time of } t \\ 0, & \textit{System is being repaired at the time of } t. \end{cases} \tag{3.3}$$

According to the definition, the instantaneous availability of a system is

$$A(t) = P\{X(t) = 1 \mid \textit{System is new at the time of } 0\}. \tag{3.4}$$

Using a total probability formula,

$$A(t) = P\Big\{X_1 > t, X(t) = 1 \mid \textit{System is new at the time of } 0\Big\}$$

$$+ P\Big\{X_1 \leq t < X_1 + Y_1, X(t) = 1 \mid \textit{System is new at the time of } 0\Big\}$$

$$+ P\Big\{X_1 + Y_1 \leq t, X(t) = 1 \mid \textit{System is new at the time of } 0\Big\}.$$

Therefore, $A(t)$ meets the renewal equation:

$$A(t) = 1 - F(t) + Q(t) * A(t). \tag{3.5}$$

Through Laplace transformation of (3.5), the instantaneous availability of a system is solved as follows $\hat{F}(s), \hat{G}(s)$, standard for Laplace transformation of $F(t)$, $G(t)$, respectively.

The average steady-state availability of a system can be obtained as follows through Tobel theorem and L'Hopital's rule of Laplace transform:

$$\tilde{A} = \frac{\mu}{\lambda + \mu}.$$

In the formula, $\lambda > 0$ and $\mu > 0$. They are parameters in (2.26) and (2.27), respectively.

Only the steady-state availability can be obtained through the aforementioned research on a renewal equation. At present, there has been some relatively mature research on this aspect. Obviously, for general probability distribution, it is very difficult to obtain an accurate analytic solution of this renewal equation. Therefore, it is important to try to find out the approximate solution of this kind of equation.

3.2 Approximate Solution of System Instantaneous Availability

When the failure time and repair time of a unit comply with general probability distribution, the instantaneous availability of single-unit continuous repairable system satisfies one renewal equation, which is second kind of Volterra integral equation.

And the successive approximation method is often used in engineering to solve the second kind of Volterra integral equation [71–73]. In References [74,75], the solution method for an integral equation is introduced; the approximation function of instantaneous availability of system with both failure time and repair time conforming to general probability distribution is solved using the successive approximation method. However, in practical engineering, the system often cannot be repaired in a timely manner after it suffers from failure due to the shortage of spare parts, environment, or man-made factors, resulting in logistic delay. Because the logistic delay time also has an important effect on equipment availability, it is necessary to take it into account.

3.2.1 Modeling

Suppose the system is composed of one single unit and its fault time X conforms to general probability distribution X; repairing the unit (system) with equipment after it is out of order and the repair time Y also conform to general probability distribution $F(t)$. After repair, the unit immediately returns to normal working status. However, there is logistic delay due to environment or personnel limitations (i.e., there is repair delay time Y between unit out of order and beginning to repair), which conforms to general probability distribution $G(t)$. After repair, the system is like one; W, $H(t)$, and X are mutually independent. For the sake of simplicity, assume the unit is a new one at the initial stage.

In order to distinguish different status of the system, the following parameters are defined:

$$\begin{cases} Z(t) = 0 & \text{\textit{System is normal at the time of} } t \\ Z(t) = 1 & \text{\textit{System is to be repaired at the time of} } t \quad t \geq 0 \\ Z(t) = 2 & \text{\textit{System is being repaired at the time of} } t \end{cases} \quad (3.6)$$

According to the availability model of a non-Markov single-unit repairable system in the previous section, the following formula is obtained:

$$A(t) = R(t) + \int_0^t A(t-u)\, dQ(u) = R(t) + Q(t) * A(t). \quad (3.7)$$

In the formula,

$$R(t) = 1 - F(t) \text{ and } Q(t) = P\{X + W + Y \leq t\} = F(t) * H(t) * G(t) = \int_0^t q(u)\, du.$$

Then, Formula 3.7 can be expressed as follows through appropriate transformation:

$$A(t) = R(t) + \int_0^t q(t-u)A(u)\,du.$$

This equation belongs to the second kind of Volterra integral equation of convolutional type, which is called a closed-loop equation by Volterra.

Because $q(t)$ is a probability density function, $M > 0$, making $q(t) \leq M$ and $\forall t \geq 0$; $R(t)$ is system reliability, making $R(t) \leq 1$ and $\forall t \geq 0$. All are real continuous functions. According to Theorem 4.1.1 in Reference [76], the integral equation (3.7) can be solved with the successive approximation method.

$$A(t) = \varphi_0(t) + \varphi_0(t) + \ldots + \ldots = \sum_{n=0}^{\infty} \varphi_n(x). \tag{3.8}$$

In that formula, $\varphi_0(t) = R(t) \leq 1$, $\varphi_n(t) = \int_0^t q(t-u)\,\varphi_{n-1}(u)\,du$. Make

$$A_n(t) = \sum_{j=0}^{n} \varphi_j(t)\, n = 0, 1, 2, \ldots. \tag{3.9}$$

Then,

$$A_{n+1}(t) = \sum_{j=0}^{n+1} \varphi_j(t) = \varphi_0(t) + \sum_{j=1}^{n+1} \varphi_j(t) = R(t) + \sum_{j=1}^{n+1} \int_0^t q(t-u)\varphi_{j-1}(u)\,du$$

$$= R(t) + \int_0^t q(t-u)\sum_{j=1}^{n+1}\varphi_{j-1}(u)\,du = R(t) + \int_0^t q(t-u)\sum_{j=0}^{n}\varphi_j(u)\,du$$

$$= R(t) + \int_0^t q(t-u)A_n(u)\,du.$$

The following iterative scheme is obtained:

$$A_0(t) = R(t) \tag{3.10}$$

$$A_{n+1}(t) = R(t) + \int_0^t q(t-u)A_n(u)\,du. \tag{3.11}$$

So, $A_n(t)$ is the approximation solution of Equation 3.7, that is, the approximate instantaneous availability of the system. From these derivations, the following formula can be obtained:

$$0 \leq A_n(t) \leq A_{n+1}(t) \leq A(t), \forall t \geq 0, \quad n = 0, 1, 2, \ldots.$$

3.2.2 Physical Significance of Instantaneous Availability Approximate Solution

According to Reference [8],

$$A(t) = P\{Z(t) = 0 \mid System\ is\ new\ at\ the\ time\ of\ 0\}. \tag{3.12}$$

In the formula, $Z(t)$ stands for system status given by (3.6).

Record $XX_i = X_i + W_i + Y_i$, $i = 1, 2, \ldots$, in which X_i, Y_i, and W_i refer to failure time, repair time, and logistic delay time, respectively, of a system in cycle i. Then, the following formula can be obtained through the total probability formula:

$$A(t) = \sum_{j=0}^{+\infty} P\left\{\sum_{i=1}^{j} XX_i \leq t < \sum_{i=0}^{j} XX_i + X_{j+1}, Z(t) = 0 \mid System\ is\ new\ at\ the\ time\ of\ 0 \right\}$$

$$+ \sum_{j=0}^{+\infty} P\left\{\sum_{i=1}^{j} XX_i + X_{j+1} \leq t < \sum_{i=0}^{j+1} XX_i, Z(t) = 0 \mid System\ is\ new\ at\ the\ time\ of\ 0 \right\}.$$

In the second term, when $\displaystyle\sum_{i=1}^{j} XX_i + X_{j+1} \leq t < \sum_{i=1}^{j+1} XX_i$, the system must be out of order at the moment of t, and it is impossible to synchronize with $Z(t) = 0$. Therefore, the second term equals zero. In the first term, make

$$B_j(t) = P\left\{\sum_{i=1}^{j} XX_i \leq t < \sum_{i=1}^{j} XX_i + X_{j+1}, Z(t) = 0 \mid System\ is\ new\ at\ the\ time\ of\ 0 \right\}. \tag{3.13}$$

Then,

$$A(t) = \sum_{j=0}^{+\infty} B_j(t).$$

In the formula,

$$B_0(t) = P\{0 \leq t < X_1, Z(t) = 0 \mid System\ is\ new\ at\ the\ time\ of\ 0\} = R(t)$$

$$B_1(t) = P\{XX_1 \leq t < XX_1 + X_2, Z(t) = 0 \mid System\ is\ new\ at\ the\ time\ of\ 0\}$$

$$= \int_0^t P\{Z(t) = 0 \mid System\ is\ new\ at\ the\ time\ of\ 0, XX_1 = u, t - u < X_2\} dp\{XX_1 \leq u\}$$

$$= \int_0^t P\{Z(t-u)=0 \,|\, \textit{System is new at the time of } 0, t-u < X_2\} dp\{XX_1 \le u\}$$

$$= \int_0^t P\{t-u < X_2\} dp\{XX_1 \le u\}$$

$$= \int_0^t R(t-u) dQ(u)$$

$$= \int_0^t B_0(t-u) dQ(u)$$

$\cdots \cdots$

$B_j(t)$

$$= P\left\{ \sum_{i=1}^{j} XX_i \le t < \sum_{i-1}^{j} XX_i + X_{j+1}, Z(t)=0 \,|\, \textit{System is new at the time of } 0 \right\}$$

$$= \int_0^t P\left\{ \sum_{i=2}^{j} XX_i \le t-u < \sum_{i=2}^{j} XX_i + X_{j+1}, Z(t)=0 \,|\, \textit{System is new at the time of } \mu \right\}$$

$$\times dp\{XX_1 \le u\}$$

$$= \int_0^t P\left\{ \sum_{i=2}^{j} XX_i \le t-u < \sum_{i=2}^{j} XX_i + X_{j+1}, Z(t-u)=0 \,|\, \textit{System is new at the time of } 0 \right\}$$

$$\times dQ(u)$$

$$= \int_0^t P\left\{ \sum_{i=1}^{j-1} XX_i \le t-u < \sum_{i=1}^{j-1} XX_i + X_j, Z(t-u)=0 \,|\, \textit{System is new at the time of } 0 \right\}$$

$$\times dQ(u)$$

$$= \int_0^t B_{j-1}(t-u) dQ(u)$$

Then, the following conclusion can be drawn:

Theorem 3.1: $A(t)$ and $A_n(t)$ are given by (3.8) and (3.9), respectively. Then,

$$A_n(t) = \sum_{j=0}^{n} B_j(t) n = 0, 1, 2, \dots \tag{3.14}$$

Prove: Using mathematical induction:

1. When $n = 0$, Formula 3.14 is obviously true.
2. Suppose Formula 3.14 is true when $n = 0, 1, \ldots, k$; then, the following content can be obtained according to (3.10) and (3.11) when $n = k + 1$:

$$A_{k+1}(t) = R(t) + \int_0^t q(t-u) A_k(u) \, du$$

$$= R(t) + \int_0^t A_k(t-u) dQ(u)$$

$$= R(t) + \int_0^t \sum_{j=0}^{k} B_j(t-u) dQ(u)$$

$$= R(t) + \sum_{j=0}^{k} \int_0^t B_j(t-u) dQ(u)$$

$$= B_0(t) + \sum_{j=0}^{k} B_{j+1}(t) = \sum_{j=0}^{k+1} B_j(t).$$

Thus, when $n = k + 1$, the proposition is true. So, Formula 3.14 is true according to (i) and (ii).

Physical meaning of $B_j(t)$: The probability that the unit failed j times and then gets repaired before t. Then, $\sum_{j=0}^{n} B_j(t)$ means the probability that unit failed at most n times and then gets repaired before t.

Physical meaning of $A_n(t)$: The system availability when the system is completely repaired and the maximum maintenance frequency is limited to n; replace the unit with spare parts immediately after it is out of order and the repair time is just the corresponding replacement time; in this case, $A_n(t)$ stands for the system availability when the spare part is n; $A_n(t)$ also can be interpreted as the reliability of a cold standby system with $n + 1$ same units.

In practical engineering, a system is often subject to limited maintenance or replacement frequency; also, the system that is repaired many times within a limited period is rejected. Therefore, the theoretical assumption "provide immediate repair after failure and can be repaired without frequency limitation" is unreasonable in some cases. Then, the availability of a real system can be understood as system instantaneous availability with repair frequency limited to N (depending on actual circumstance). According to the physical meaning of approximate instantaneous availability obtained in this chapter, it is relatively reasonable to describe the availability of a real system with approximate instantaneous availability $A_N(t)$.

3.2.3 Error Analysis of System Approximate Availability

Definition 3.1: $A(t)$ and $A_n(t)$ are given by (3.8) and (3.9), respectively. Then,

$$\delta_n(t) = A(t) - A_n(t) \; t \geq 0 \tag{3.15}$$

is the approximate availability error function.

The upper and lower bounds of approximate availability error (3.15) are analyzed as follows. According to previous conclusions,

$$A(t) = \sum_{j=0}^{\infty} B_j(t) = \sum_{j=0}^{n} B_j(t) + \sum_{j=n+1}^{\infty} B_j(t) = A_n(t) + \sum_{j=n+1}^{\infty} B_j(t).$$

As per the definition (3.13) of $B_j(t)$, $\delta_n(t) = \sum_{j=n+1}^{\infty} B_j(t) \geq 0$ is obtained:

$$\delta_n(t) = \sum_{j=n+1}^{\infty} B_j(t) = \sum_{j=n+1}^{\infty} P\left\{ \sum_{i=1}^{j} XX_i \leq t < \sum_{i=1}^{j} XX_i + X_{j+1} \right\}$$

$$= \sum_{j=n+1}^{\infty} \left(P\left\{ \sum_{i=1}^{j} XX_i \leq t \right\} - P\left\{ \sum_{i=1}^{j} XX_i + X_{j+1} \leq t \right\} \right)$$

$$= P\left\{ \sum_{i=1}^{n+1} XX_i \leq t \right\} - \sum_{j=n+1}^{\infty} \left(P\left\{ \sum_{i=1}^{j} XX_i + X_{j+1} \leq t \right\} - P\left\{ \sum_{i=1}^{j+1} XX_i \leq t \right\} \right).$$

For arbitrary j,

$$\left\{ \sum_{i=1}^{j} XX_i + X_{j+1} \leq t \right\} \supseteq \left\{ \sum_{i=1}^{j+1} XX_i \leq t \right\}.$$

Thus,

$$P\left\{ \sum_{i=1}^{j} XX_i + X_{j+1} \leq t \right\} - P\left\{ \sum_{i=1}^{j+1} XX_i \leq t \right\} \geq 0$$

$$\sum_{j=n+1}^{\infty} \left(P\left\{ \sum_{i=1}^{j} XX_i + X_{j+1} \leq t \right\} - P\left\{ \sum_{i=1}^{j+1} XX_i \leq t \right\} \right) \geq 0.$$

Then, the upper and lower bounds of approximate availability error are

$$0 \le \delta_n(t) \le P\left\{\sum_{i=1}^{n+1} XX_i \le t\right\}. \tag{3.16}$$

So then, the error of system approximate availability is obtained through the successive approximation method. Make

$$\bar{\delta}_N(t) = P\left\{\sum_{i=1}^{N+1} XX_i \le t\right\}.$$

Then, $\bar{\delta}_N(t)$ is an upper boundary of system approximate availability error with approximation number $= N$ at the time of t, which is called the error bound function of system approximate availability. In this case, the research on system approximate availability error can be transformed to study error bound function $\bar{\delta}_N(t)$.

Through this analysis, the following properties related to error bound function $\bar{\delta}_N(t)$ can be obtained:

1. For arbitrary N, $\bar{\delta}_N(t)$ is monotone increasing function of t and satisfies $\bar{\delta}_N(0) = 0$ and $\bar{\delta}_N(\infty) = 1$.

2. For arbitrary t, $\bar{\delta}_N(t)$ is monotone decreasing sequence of N.

According to property (1) of $\bar{\delta}_N(t)$, $\left\{\bar{\delta}_N(T) \le \varepsilon_0\right\} \Rightarrow \left\{\bar{\delta}_N(t) \le \varepsilon_0, \forall 0 \le t \le T\right\}$ and $N = 0, 1, 2, \ldots$, through which the following conclusions can be drawn:

Theorem 3.2: When the time interval for research is $[0, T]$ and the accuracy requirement of instantaneous availability is ε_0, if there is natural number N and $\bar{\delta}_N(T) \le \varepsilon_0$, the instantaneous availability of the original system can be approximately replaced with system approximate availability $A_N(t)$.

Note: To facilitate the calculation, in practical engineering, it is only necessary to use $A_n(t)$ to approximate system instantaneous availability, in which,

$$n = \inf\left\{N \mid \bar{\delta}_N(T) \le \varepsilon_0\right\}.$$

3.2.4 Numerical Example

An example is given as follows to describe the reasonableness of replacing system instantaneous availability with system approximate availability.

For example, suppose that the failure time, logistic delay time, and repair time of some single-unit repairable system conform to Weibull distributions $W(8, 0.4)$, $W(1.5, 0.95)$, and $W(1.2, 0.99)$, respectively; all such time is measured with days. Objectives are to solve the availability $\tilde{A}(t)$ of the system and to meet the accuracy requirement (i.e., the error of any time within 20 days shall be less than 0.01). Such objectives can be expressed with the following mathematical expression:

$$\left|\tilde{A}(t) - A(t)\right| \leq 0.01 \ \forall t \in [0, 20].$$

First, determine the approximation number n according to accuracy requirement:

$$n = \inf\left\{N \mid \bar{\delta}_N(20) \leq 0.01\right\}.$$

Table 3.1 shows corresponding error upper bounds of different approximation numbers. To ensure intuitional instruction, in Figure 3.2, error bound functions of system approximate availability for approximation number is equal to 2, 4, and 5 are presented. From the figure, it can be seen that in time interval [0, 20], the error bound function $\bar{\delta}_5(t)$ always has a very small value. According to Figure 3.2 and Table 3.1, it can be seen that $n = 5$.

Therefore, the system approximate availability $A_5(t)$ with approximation number = 5 can meet the accuracy requirement. Figure 3.3 shows the comparison of system approximate availability and actual instantaneous availability when the approximation number is equal to 2, 4, and 5. It can be seen that in time interval [0, 8], the system approximate availability $A_2(t)$ can well approach system instantaneous availability; in time interval [0, 15], the system approximate availability $A_4(t)$ can well approach system instantaneous availability; in time interval [0, 20], the system approximate availability $A_5(t)$ can well approach system instantaneous availability, but $A_4(t)$ is not satisfied.

This content presents the approximate solution of instantaneous availability of a single-unit repairable system with failure time, repair time, and logistic delay time all subject to general probability distribution. The approximate system instantaneous

Table 3.1 Error Bounds of System Approximate Availability with Different Approximation Numbers

Approximation Number N	Error Upper Bound $\bar{\delta}_N(20)$
0	0.9992
1	0.9984
2	0.9967
3	0.9060
4	0.3123
5	0.0099

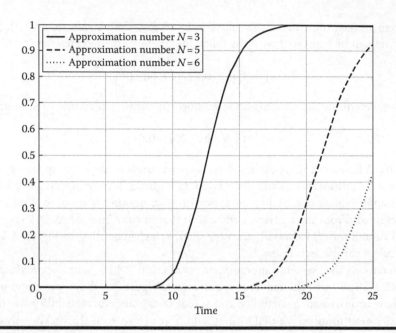

Figure 3.2   Error bound functions of system approximate availability with different approximation numbers.

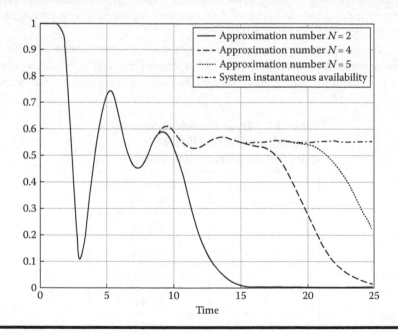

Figure 3.3   Curves of system approximate availability with different approximation numbers.

availability can be obtained through the aforementioned method. Also, within a specific time interval, the error can be controlled within a specified range only if an appropriate approximation number is selected. This is a new method for solving system instantaneous availability. This method has definite physical meaning and is also commonly used. However, as for fluctuation research, this approximation method still has many problems, such as constant selection of approximation numbers according to research content, research on consistency of approximate solution and actual fluctuation law, and so on—all of which constrain our application of approximate solution. Furthermore, strictly speaking, this approximate solution is not am analytic solution. Therefore, the basis for problem analysis through direct mathematical analysis is not very solid, which also has an effect on the accuracy of fluctuation analysis, thus directly affecting the value of theoretical research and engineering application.

3.3 Instantaneous Availability Model of Single-Unit Repairable System with Partial Repair Time Negligible

3.3.1 Problem Proposed

Another practical problem is also found when researching the instantaneous availability model under general probability distribution (i.e., repair system immediately when it fails) in practical engineering. However, if the failure can be repaired within a very short time and has no effect on the normal operation of a system, the repair time can be neglected. In this case, we still think the system always keeps running. This case sounds very much like that of the particle channel. If we open and close the channel constantly, the time interval between such opening and closing is so short that the switching cannot be recorded. At this point, we do not think that there is switching (this is a so-called time omission) [77–79]. The correct establishment of such a system and calculation of its availability can better facilitate the solving of some practical problems that conform to maintenance theory.

At present, some domestic and international scholars are also studying this problem. However, many studies refer to the research on the Markov repairable system [13,80,81] in which the Reference [13] presents the establishment of a new model with partial repair time negligible similar to the ion-channel model for a single-unit Markov repairable system. It supposes that if the repair time is short enough (less than some critical value) and has little effect on system operation, such repair time can be deleted from down-time. When the failure rate and repair rate of a system are constant, the system status satisfies the Markov process, and system reliability and availability analytical form can be accurately obtained. In this case, the research is clear and concise and easy to carry out, which are also important reasons for why the failure rate and repair rate are supposed to be constant at the time of product research (especially with an electronic product).

However, in practice, the failure rate of most products varies with time during the entire service life. For example, the decline failure rate curve, bathtub failure rate curve, roller coaster failure rate curve, and zigzag failure rate curve all have been applied in many practical projects [14–16]. Therefore, the Markov repairable system model is not always reasonable in many cases. For this reason, it is necessary to extend the research range to a non-Markov repairable system [16,82]. Suppose the failure time and repair time of a system are both subject to general probability distribution, a similar non-Markov single-unit repairable system model is given, and the instantaneous availability of the model is obtained.

3.3.2 Mathematical Model

First, suppose

1. System starts to run when time = 0.
2. System is repaired immediately after it is out of order.
3. All normal time and repair time during this period are mutually independent.
4. System is completely repaired (i.e., system is as good as new after repair).
5. X_j and Y_j stand for time between normal status and time between failure status jth, respectively, both of which have respective independent identical distribution and the distribution functions of which refer to $F(t)$ and $G(t)$ with $j = 1, 2, \ldots$, respectively.

For convenience, the normal status of the system is labeled as 0 and the failure status as 1. Then, the status space can be expressed as {0, 1}. The system status is labeled as $Z_0(t)$.

$$Z_0(t) = \begin{cases} 0 & \textit{System normal at the time of } t \\ 1 & \textit{System failed at the time of } t. \end{cases}$$

In the following section, a new model is outlined to describe some properties in practical engineering. First, a critical value $\tau(\tau \geq 0)$ is given; in each repair course, if the repair time is less than τ and has little effect on system operation, it is thought that the system still keeps working normally during this period; otherwise, the system is in failure status, as shown in Figure 3.4.

Then, a new model can be obtained based on the original system:

1. The new model has two kinds of status: normal and faulted.
2. If the original system operates normally, the new model also has normal operation status.

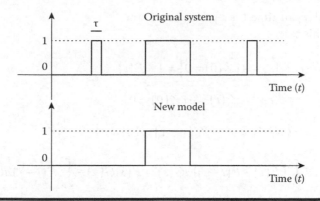

Figure 3.4 Comparison of original system and new model.

3. If original system is out of order and the repair time is shorter than τ, it is thought that the new model operates normally and the failure did not happen.
4. If original system is out of order and the repair time is longer than τ, it is thought that the new model is faulted.

Then, a random process $\{Z_1(t), t \geq 0\}$ can be used to describe the status of the new model in which

$$Z_1(t) = \begin{cases} 0 & \text{New model normal at the time of } t \\ 1 & \text{New model failed at the time of } t. \end{cases}$$

In following sections, a study of instantaneous availability of original system and new model is carried out.

3.3.3 Establishment of Instantaneous Availability Model

Make $A(t)$ be instantaneous availability of original system and make it satisfy renewal equation (3.5); make $A_1(t)$ stand for instantaneous availability of new model, then,

$$A_1(t) = P\{X_1 > t, Z_1(t) = 0 \mid Z(0) = 0\} + P\{X_1 \leq t < X_1 + Y_1, Z_1(t) = 0 \mid Z(0) = 0\} + P\{X_1 + Y_1 \leq t, Z_1(t) = 0 \mid Z(0) = 0\}.$$

According to the selection of critical repair time of the Markov repairable system in Reference [13], in this chapter, the critical repair time is discussed under two kinds of circumstances.

1. Critical repair time τ is a positive constant.
   In this case,

$$P\{X_1 > t, Z_1(t) = 0 \mid Z(0) = 0\} = 1 - F(t)$$

$$P\{X_1 \le t < X_1 + Y_1, Z_1(t) = 0 \mid Z(0) = 0\}$$

$$= P\{X_1 \le t < X_1 + Y_1, Y_1 < \tau \mid Z(0) = 0\}$$

$$= \int_0^\tau (F(t) - F(t-s)) dG(s) = F(t)G(\tau) - \int_0^\tau F(t-s) dG(s)$$

$$P\{X_1 + Y_1 \le t, Z_1(t) = 0 \mid Z(0) = 0\}$$

$$= \int_0^t P\{Z_1(t) = 0 \mid Z(0) = 0, X_1 + Y_1 = u\} dP\{X_1 + Y_1 \le u\}$$

$$= \int_0^t P\{Z_1(t) = 0 \mid Z(u) = 0\} dP\{X_1 + Y_1 \le u\}$$

$$= \int_0^t P\{Z_1(t-u) = 0 \mid Z(0) = 0\} dQ(u)$$

$$= \int_0^t A_1(t-u) dQ(u) = Q(t) * A_1(t).$$

Then, the instantaneous availability of the new model satisfies

$$A_1(t) = 1 - F(t) + F(t)G(\tau) - \int_0^\tau F(t-s) dG(s) + Q(t) * A_1(t). \quad (3.17)$$

2. Critical repair time τ is a nonnegative random variable.
   Suppose the distribution function of τ is $H(t)$, like the conclusion of part (1), there is

$$P\{X_1 > t, Z_1(t) = 0 \mid Z(0) = 0\} = 1 - F(t)$$

$$P\{X_1 \le t < X_1 + Y_1, Z_1(t) = 0 \mid Z(0) = 0\}$$

$$= P\{X_1 \le t < X_1 + Y_1, Y_1 < \tau \mid Z(0) = 0\}$$

$$= \int_0^t f(s) ds \int_{t-s}^{+\infty} g(u)(1 - H(u)) du$$

$$P\{X_1+Y_1 \le t, Z_1(t)=0 \mid Z(0)=0\}$$

$$= \int_0^t P\{Z_1(t)=0 \mid Z(0)=0, X_1+Y_1=u\} dP\{X_1+Y_1 \le u\}$$

$$= \int_0^t P\{Z_1(t)=0 \mid Z(u)=0\} dP\{X_1+Y_1 \le u\}$$

$$= \int_0^t P\{Z_1(t-u)=0 \mid Z(0)=0\} dQ(u)$$

$$= \int_0^t A_1(t-u) dQ(u) = Q(t) * A_1(t).$$

So, the instantaneous availability of the new model satisfies

$$A_1(t)=1-F(t)+\int_0^t f(s)ds \int_{t-s}^{+\infty} g(u)(1-H(u))du + Q(t) * A_1(t). \quad (3.18)$$

3.3.4 Numerical Example

In this section, numerical examples are given to describe the model in this chapter under two kinds of circumstances: that is, τ equals the positive constant and non-negative random variable, respectively; when τ equals a nonnegative random variable, research is carried out when the system is subject to exponential distribution and Weibull distribution, respectively, through which instantaneous availabilities of the new model and original system are obtained. Then, we analyze and compare them.

Suppose X_j is subject to Weibull distribution $W(\alpha_1, \lambda_1; t)$ and Y_j to Weibull distribution $W(\alpha_2, \lambda_2; t)$ with $j = 1, 2, \ldots$ and make $\alpha_1 = 1.4$, $\lambda_1 = 0.6$, $\alpha_2 = 1.5$, and $\lambda_2 = 1.5$, then

$$F(t)=1-\exp\left[-(0.6t)^{1.4}\right], \quad G(t)=1-\exp\left[-(1.5t)^{1.5}\right], \quad t \ge 0.$$

Obviously, the system is a non-Markov system. Its instantaneous availability is shown in Figure 3.5.

In the following paragraphs, several different circumstances of the new model are presented.

Example 1

When $\tau = 0.4$, the instantaneous availability of the new model can be obtained according to Formula 3.17, as shown in Figure 3.6.

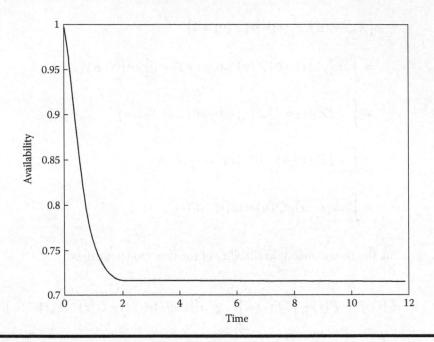

Figure 3.5   Instantaneous availability curve of original.

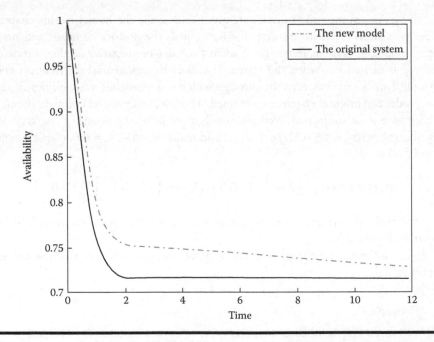

Figure 3.6   Instantaneous availability curves of new model and original system when τ = 0.4.

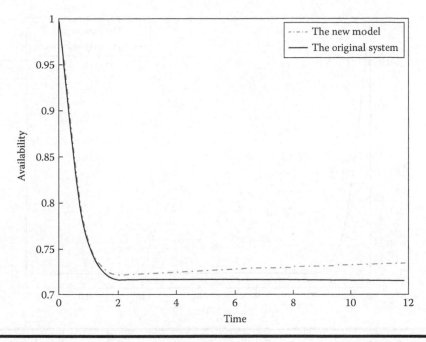

Figure 3.7 **Instantaneous availability models of original system and new model when τ is subject to exponential distribution.**

Example 2

When τ is subject to the exponential distribution of $\lambda_3 = 0.4$, the instantaneous availability of the new model can be obtained as per Formula 3.18, as shown in Figure 3.7.

Example 3

When τ is subject to Weibull distribution $W(\alpha_4, \lambda_4; t)$, that is,

$$H(t) = 1 - \exp\left(-\left(\lambda_4 t\right)^{\alpha_4}\right), \quad t \geq 0.$$

Make $\alpha_4 = 1.5$ and $\lambda_4 = 1.1$. Then, the instantaneous availability of the new model can be obtained according to Formula 3.5, as shown in Figure 3.8.

From Figures 3.6, 3.7, and 3.8, it can be intuitively seen that no matter whether τ is a positive constant or a nonnegative random variable, the instantaneous availability of the new model is always higher than that of the original system, which conforms to the actual situation. For the approximate solution of this kind of problem, the research focuses on establishment of a renewable equation for approximation. From the view of effectiveness of a numerical calculation, it has certain engineering practicability. However, for research on fluctuation, the proper approximate analytic solution still has not been found, which reflects the same problem as with the previous section.

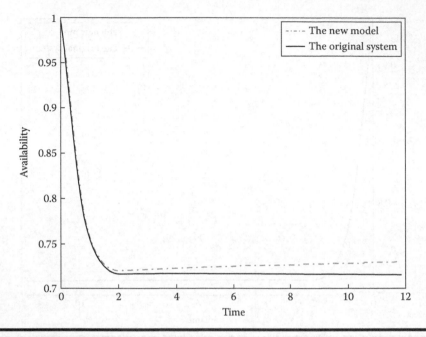

Figure 3.8 **Instantaneous availability curves of new model and original system when τ is subject to Weibull distribution.**

3.4 Summary

In this chapter, the integral equation of instantaneous availability of a single-unit continuous repairable system with failure time, repair time, and logistic delay time all subject to general probability distribution is solved through a successive approximation method thus obtaining the approximate solution of system instantaneous availability. In addition, the physical meaning of system approximate availability is researched and error analysis of approximate solution is carried out. Also, a new model is also established for a non-Markov single-unit repairable system with a short repair time having little effect on normal operation of a system for some practical situations; its instantaneous availability is also obtained. This method can be further promoted for systems of other structures. These two kinds of approximate solution have some errors, which can be controlled accordingly. However, from the view of research on fluctuation of instantaneous availability, the approximate solution has following main problem: From a comparison of numerical solutions, the fluctuation law of approximate solution cannot remain consistent with that of the numerical solution of the equation. Even if an adjustment of the approximation number can improve this problem, the problem reflects the principles that these two solutions are not completely the same. Obviously, such an analytic solution neither meets the requirement of approximate accuracy nor facilitates research from the view of mathematical analysis. All of these problems will affect the research on principles of instantaneous availability fluctuation.

Therefore, from the view of engineering, it is necessary to continue to seek a new problem-solving method. There are two main approaches for this: (1) directly carrying out a numerical solution for renewal equation, which is a traditional method when the analytic solution cannot be found. However, it is necessary to pay much attention to the fact that convolution operation is required for solution of the renewal equation, while convolution operation is very time consuming during numerical solution; (2) directly discretizing a continuous problem before equation establishment and then establishing a new discrete equation to research a numerical solution. In this case, people often avoid complex calculations to greatly simplify calculations, thus laying a foundation for application of such methods in engineering. Meanwhile, the direct discretization of a problem can facilitate the use of a uniform modeling method and calculation method, thus facilitating problem solving from a methodology aspect.

Chapter 4

Instantaneous Availability Model of Discrete-Time System under General Probability Distribution

In the previous chapter, detailed research on the availability equation and its approximate solution of continuous system under general probability distribution is carried out, which is significant for research on instantaneous availability fluctuation to a certain degree. Through analysis, it is found that there is still a great limitation in the aforementioned research results, that is, the approximate solution does not really equal the analytic solution, and the error of approximate solution has an influence on analysis of fluctuation. Nevertheless, in order to better study instantaneous availability fluctuation, it is necessary to seek a new method for solving the instantaneous availability equation. As everyone knows, there is no very effective method for determining the analytic solution and approximate solution of instantaneous availability equation. This chapter focuses on trying to solve the equation using a numerical computation method.

The numerical computation method mainly can be divided into two sections: (1) directly carrying out the numerical solution for the instantaneous availability equation established in previous chapter (this solution refers to complex convolution calculation, and because it is different to control the calculation error and the calculation amount is very large, this method is not very convenient for application in engineering); and (2) directly establishing instantaneous availability equation of discrete time and then carrying out numerical calculations to solve the equation. For this kind of method, the calculation amount is relatively small and its physical significance is clear. So, it is a relatively convenient application for use in engineering. In this chapter, the discrete-time instantaneous availability equation of a single-unit repairable system is first established to research instantaneous availability models of a single-unit repairable system with logistic delay time considered, a single-unit repairable system with preventive maintenance interval time considered, and multi-unit series and multi-unit parallel repairable systems, so as to lay the foundation for numerical calculation and analysis of instantaneous availability.

4.1 Single-Unit Repairable System under Discrete Time

4.1.1 Modeling

Suppose the system is composed of a single unit, and its failure time X complies with general discrete distribution:

$$p_k = P\{X = k\} \quad k = 0, 1, 2, \ldots$$

If the system is repaired immediately after it is out of order, the system is as good as new after repairing. Therefore, the repair time Y complies with general discrete distribution:

$$q_k = P\{Y = k\} \quad k = 0, 1, 2, \ldots$$

In order to distinguish different status of the system, the following parameters are defined:

$$\begin{cases} Z(k) = 0 & \textit{System normal at the time of } K \\ Z(k) = 1 & \textit{System failed at the time of } K \end{cases} \quad k = 0, 1, 2, \ldots$$

In order to not lose generality, suppose the unit is new at the initial stage, that is, $P\{Z(0) = 0\} = 1$.

The function of system failure rate is

$$\lambda(k) = P\{X = k | X \geq k\} \quad k = 0, 1, 2, \ldots$$

The function of system repair rate is

$$\mu(k) = P\{Y = k | Y \ge k\} \quad k = 0, 1, 2, \dots$$

Make $P_0(k, j)$ stand for the probability that the system has been in 0 status for j time at the time of k and $P_1(k, j)$ stand for the probability that system has been in 1 status for j time at the time of k; that is, when $j = 0, 1, \dots, k - 1$,

$$\begin{cases} P_0\ (k, j) = P\{Z(k) = Z\ (k-1) = \dots = Z(k-j) = 0, Z(k-j-1) = 1\} \\ P_1\ (k, j) = P\{Z(k) = Z\ (k-1) = \dots = Z(k-j) = 1, Z(k-j-1) = 0\}. \end{cases}$$

When $j = k$,

$$\begin{cases} P_0(k,k) = P\{Z(k) = Z(k-1) = \dots = Z(0) = 0\} \\ P_1(k,k) = P\{Z(k) = Z(k-1) = \dots = Z(0) = 1\}. \end{cases}$$

When $k < j$, $P_0(k, j) = 0$ and $P_1(k, j) = 0$.

When $j = 0, 1, \dots, k - 1$, probability analysis is used to obtain the following content:

$$P_0\left(k+1, j+1\right) = P\{Z(k+1) = Z(k) = \dots = Z(k-j) = 0, Z(k-j-1) = 1\}$$

$$= P\{Z(k) = Z(k-1) = \dots = Z(k-j) = 0, Z(k-j-1) = 1\} \times$$

$$P\{[Z(k+1) = 0] | [Z(k) = Z(k-1) = \dots = Z(k-j) = 0, Z(k-j-1) = 1]\}$$

$$= P_0\left(k, j\right)\left(1 - \lambda\left(j\right)\right)$$

$$P_1\left(k+1, j+1\right) = P\{Z(k+1) = Z(k) = \dots = Z(k-j) = 1, Z(k-j-1) = 0\}$$

$$= P\{Z(k) = Z(k-1) = \dots = Z(k-j) = 1, Z(k-j-1) = 0\} \times$$

$$P\{[Z(k+1) = 1] | [Z(k) = Z(k-1) = \dots = Z(k-j) = 1, Z(k-j-1) = 0]\}$$

$$= P_1\left(k, j\right)\left(1 - \mu\left(j\right)\right).$$

In addition,

$$P_0(k+1, k+1) = P\{Z(k+1) = Z(k) = \ldots Z(0) = 0\}$$
$$= P\{Z(k) = Z(k-1) = \ldots = Z(0) = 0\}$$
$$\times P\{[Z(k+1) = 0] | [Z(k) = Z(k-1) = \ldots = Z(0) = 0]\}$$
$$= P_0(k, k)(1 - \lambda(k))$$

$$P_1(k+1, k+1) = P\{Z(k+1) = Z(k) = \ldots = Z(0) = 1\}$$
$$= P\{Z(k) = Z(k-1) = \ldots = Z(0) = 1\}$$
$$\times P\{[Z(k+1) = 1] | [Z(k) = Z(k-1) = \ldots = Z(0) = 1]\}$$
$$= P_1(k, k)(1 - \mu(k))$$

$$P_0(k+1, 0) = P\{Z(k+1) = 0, Z(k) = 1\}$$
$$= \sum_{j=0}^{k} P\{Z(k+1) = 0, Z(k) = Z(k-1) = \ldots = Z(k-j) = 1, Z(k-j-1) = 0\}$$
$$= \sum_{j=0}^{k} P\{Z(k) = Z(k-1) = \ldots = Z(k-j) = 1, Z(k-j-1) = 0\}$$
$$\times P\{[Z(k+1) = 0] | [Z(k) = Z(k-1) = \ldots Z(k-j) = 1, Z(k-j-1) = 0]\}$$
$$= \sum_{j=0}^{k} P_1(k, j)\mu(j)$$

$$P_1(k+1, 0) = P\{Z(k+1) = 1, Z(k) = 0\}$$
$$= \sum_{j=0}^{k} P\{Z(k+1) = 1, Z(k) = Z(k-1) = \ldots = Z(k-j) = 0, Z(k-j-1) = 1\}$$
$$= \sum_{j=0}^{k} P\{Z(k) = Z(k-1) = \ldots = Z(k-j) = 0, Z(k-j-1) = 1\}$$
$$\times P\{[Z(k+1) = 1] | [Z(k) = Z(k-1) = \ldots = Z(k-j) = 0, Z(k-j-1) = 1]\}$$
$$= \sum_{j=0}^{k} P_1(k, j)\mu(j).$$

Composite results are as follows:

$$
\begin{cases}
P_0(k+1,j+1) = P_0(k,j)(1-\lambda(j)) \\
P_1(k+1,j+1) = P_1(k,j)(1-\mu(j)) \\
P_0(k+1,0) = \displaystyle\sum_{j=0}^{k} P_1(k,j)\mu(j) \\
P_1(k+1,0) = \displaystyle\sum_{j=0}^{k} P_0(k,j)\lambda(j) \\
P_0(0,0) = 1, P_1(0,0) = 0.
\end{cases}
$$

4.1.2 Main Results

Using the model in Section 4.1.1, the corresponding reliability index is given.

4.1.2.1 System Reliability

Definition 4.1: The random variable X of a positive integer is used to describe the service life of a system or unit. The corresponding distribution sequence is as follows:

$$
p_k = P\{X = k\} \quad k = 0, 1, 2, \ldots
$$

The probability that the system is always normal (not failed) before k (including k) is

$$
R(k) = P\{X > k\} = \sum_{l=k+1}^{\infty} p_l.
$$

It is defined as the reliability function or reliability degree of the system.

Definition 4.2: The mean time to the first failure of system is defined as

$$
T_M = E(X) = \sum_{l=k+1}^{\infty} l p_l.
$$

Theorem 4.1: The model in Section 4.1.1 is used to establish the following two formulas:

$$
R(k) = P_0(k, k)
$$

$$
T_M = \lambda(0) \sum_{j=1}^{\infty} j P_0(j-1, j-1).
$$

Prove: The following formula can be obtained according to definitions of reliability and mean time to the first failure of system:

$$R(k) = P\{X > k\} = P\{Z(k) = Z(k-1) = \ldots = Z(0) = 0\}.$$

In other words, $R(k) = P_0(k, k)$.

$$T_M = E(X) = \sum_{j=1}^{\infty} jP\{Z(0) = \ldots = Z(j-1) = 0, Z(j) = 1\}$$

$$= \sum_{j=1}^{\infty} jP\{Z(0) = \ldots = Z(j-1) = 0\} P\{Z(j=1) | [Z(0) = \ldots = Z(j-1) = 0]\}$$

$$= \lambda(0) \sum_{j=1}^{\infty} jP_0(j-1, j-1)$$

4.1.2.2 System Availability

Definition 4.3: The probability that the system is in normal status at the time of k is defined as instantaneous availability; that is,

$$A(k) = P\{X(k) = 0\}.$$

Theorem 4.2: The model established in Section 4.1.1 is used to obtain following formula:

$$A(k) = \sum_{j=0}^{k} P_0(k, j).$$

Prove: Use the probability analysis method according to the definition of system availability:

$$A(k) = P\{Z(k) = 0\}$$

$$= P\{Z(k) = 0, Z(k-1) = 1\} + P\{Z(k) = Z(k-1) = 0, Z(k-2) = 1\} + \ldots$$

$$+ P\{Z(k) = Z(k-1) = \ldots = Z(0) = 0\}$$

$$= P_0(k, 0) + P_0(k, 1) + \ldots + P_0(k, k) = \sum_{j=0}^{k} P_0(k, j)$$

4.1.2.3 Mean Failure Frequency

In Reference [66], the mean failure frequency $M(k) = E[N(k)]$ of system within $(0, k)$ is defined. The following formula is obtained according to the model in Section 4.1.1:

$$M(k) = P_0(1,1) M(k-1) + \sum_{l=1}^{k-1} \{P_1(l, l-1) P_0(l+1, 0)[1 + M(k-l)]\} + P_1(k, k-1).$$

Prove:

$$M(k) = E[N(k)] = P\{Z(1) = Z(0) = 0\}E\{N(k) \mid Z(1) = Z(0) = 0\}$$

$$+ \sum_{l=1}^{k-1} \left\{ \begin{array}{l} P\{Z(l+1) = 0, Z(l) = \ldots = Z(1) = 1, Z(0) = 0\} \\ \times E\{N(k) \mid Z(l+1) = 0, Z(l) = \ldots = Z(1) = 1, Z(0) = 0\} \end{array} \right\}$$

$$+ P\{Z(k) = \ldots = Z(1) = 1, Z(0) = 0\}$$

$$= P_0(1,1)E[N(k-1)]$$

$$+ \sum_{l=1}^{k-1} \left\{ \begin{array}{l} P\{Z(l) = \ldots = Z(1) = 1, Z(0) = 0\} \\ \times P\{Z(l+1) = 0 \mid Z(l) = \ldots = Z(1) = 1, Z(0) = 0\} \\ \times [1 + M(k-l-1)] \end{array} \right\}$$

$$= P_0(1,1)M(k-1) + \sum_{l=1}^{k-1} \left\{ P_1(l, l-1)\mu(l-1)[1 + M(k-l-1)] \right\} + P_1(k, k-1)$$

4.1.3 Numerical Example

Several numerical examples are given here. For simplicity, only reliability and availability are studied.

Example 1

The failure time X and repair time Y of the system unit are both subject to geometric distribution:

$$p_k = P\{X = k\} = p(1-p)^k, \quad q_k = P\{Y = k\} = q(1-q)^k \quad k = 0, 1, 2, \ldots$$

The corresponding failure rate and repair rate of the system are both constant. When $p = 0.3$ and $q = 0.8$, the failure rate and repair rate are $\lambda = 0.3$ and $\lambda = 0.3$, respectively. The reliability and availability of the system are obtained according to models as shown in Figure 4.1.

Example 2

Failure time X of unit is subject to discrete Weibull distribution:

$$p_k = P\{X = k\} = p^{k^\alpha} - p^{(k+1)^\alpha}, \quad k = 0, 1, 2, \ldots$$

Repair time Y is subject to discrete Weibull distribution:

$$q_k = P\{Y = k\} = q^{k^\beta} - q^{(k+1)^\beta}, \quad k = 0, 1, 2, \ldots$$

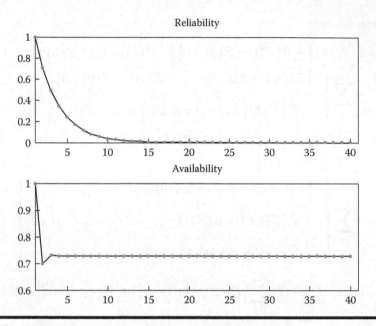

Figure 4.1   Reliability and availability of system when failure time and repair time of unit are both subject to geometric distribution.

The corresponding failure rate of system is

$$\lambda(k)=1-p^{(k+1)^\alpha-k^\alpha},\quad k=0,1,2,\ldots$$

The repair rate is

$$\mu(k)=1-q^{(k+1)^\beta-k^\beta},\quad k=0,1,2,\ldots$$

when $p = 0.4$, $\alpha = 1.2$, $q = 0.7$ and $\beta = 0.8$.

The reliability and availability of system is obtained according to models in previous two parts, as shown in Figure 4.2.

Example 3

The failure rate and repair rate of the system, respectively, are as follows:

$$\lambda(k)=\begin{cases} 16^k e^{-16}/k!\ k\le80 \\ 0.07 \qquad k>80 \end{cases}$$

$$\mu(k)=\begin{cases} 17.5^k e^{-17.5}/k!\ k\le80 \\ 0.06 \qquad k>80. \end{cases}$$

The reliability and availability of system is obtained according to models shown in Figure 4.3.

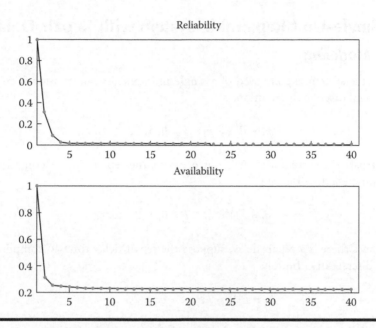

Figure 4.2 **Reliability and availability of system when failure time and repair time of unit are subject to discrete Weibull distribution.**

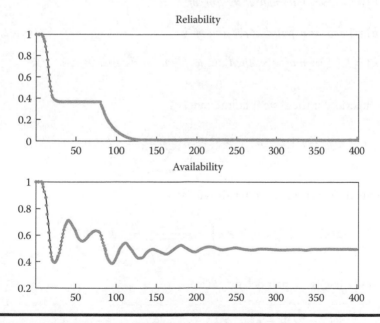

Figure 4.3 **Reliability and availability of system with failure rate and repair rate functions given directly.**

4.2 Single-Unit Repairable System with Repair Delay

4.2.1 Modeling

Suppose the system is composed of a single unit, and its failure time X complies with general discrete distribution:

$$p_k = P\{X = k\} \quad k = 0, 1, 2, \ldots$$

To repair the system after it is out of order, the repair time Y complies with general discrete distribution:

$$q_k = P\{Y = k\} \quad k = 0, 1, 2, \ldots$$

Because there is a repair delay, suppose the repair delay time W complies with general discrete distribution:

$$e_k = P\{W = k\} \quad k = 0, 1, 2, \ldots$$

In order to distinguish different status of the system, the following parameters are defined:

$$\begin{cases} Z(k) = 0 & \textit{System normal at the time of } k \\ Z(k) = 1 & \textit{System failed at the time of } k \\ Z(k) = 2 & \textit{System to be repaired due to failure at the time of } k \end{cases} \quad k = 0, 1, 2, \ldots$$

The function of the system failure rate is

$$\lambda(k) = P\{X = k \mid X \geq k\} = \frac{p_k}{1 - \sum_{i=0}^{k-1} p_i} \quad k = 0, 1, 2, \ldots$$

The function of the system repair rate is

$$\mu(k) = P\{Y = k \mid Y \geq k\} = \frac{q_k}{1 - \sum_{i=0}^{k-1} q_i} \quad k = 0, 1, 2, \ldots$$

Because there is a repair delay,

$$\rho(k) = P\{W = k \mid W \geq k\} = \frac{e_k}{1 - \sum_{i=0}^{k-1} e_i} \quad k = 0, 1, 2, \ldots$$

Therefore, $P_0(k, j)$ stands for the probability that system has been in 0 status for j time at the time of k; $P_1(k, j)$ stands for the probability that system has been in 1 status for j time at the time of k; and $P_2(k, j)$ stands for the probability that system has been in 2 status for j time at the time of k, i.e., when $j = 0, 1, \ldots, k - 1$,

$$
\left[
\begin{array}{l}
P_0(k,j) = P\left\{ \begin{array}{l} Z(k) = \ldots = Z(k-j) = 0 \\ Z(k-j-1) = 1 \end{array} \right\} \\[2ex]
P_1(k,j) = P\left\{ \begin{array}{l} Z(k) = \ldots = Z(k-j) = 1 \\ Z(k-j-1) = 2 \end{array} \right\} \\[2ex]
P_2(k,j) = P\left\{ \begin{array}{l} Z(k) = \ldots = Z(k-j) = 2 \\ Z(k-j-1) = 0 \end{array} \right\}
\end{array}
\right.
$$

$$
\left\{
\begin{array}{l}
P_0(k,k) = P\{Z(k) = Z(0) = 0\} \\[1ex]
P_1(k,k) = P\{Z(k) = Z(0) = 1\} \\[1ex]
P_2(k,k) = P\{Z(k) = Z(0) = 2\}
\end{array}
\right.
$$

when $j = k$ and $k < j$,

$$
P_0(k, j) = 0, \quad P_1(k, j) = 0, \quad P_2(k, j) = 0.
$$

The mathematical model of the repairable system can be obtained as follows through probability analysis:

$$
\left\{
\begin{array}{l}
P(k+1, j+1) = H(j)P(k, j) \\[2ex]
P(k+1, 0) = \displaystyle\sum_{j=0}^{+\infty} B(j)P(k, j).
\end{array}
\right.
$$

In the formula,

$$
P(k, j) = (P_0(k, j), P_1(k, j), P_2(k, j))^T
$$

$$H(j) = \begin{bmatrix} 1-\lambda(j) & 0 & 0 \\ 0 & 1-\rho(j) & 0 \\ 0 & 0 & 1-\mu(j) \end{bmatrix}$$

$$B(j) = \begin{pmatrix} 0 & 0 & \mu(j) \\ \lambda(j) & 0 & 0 \\ 0 & \rho(j) & 0 \end{pmatrix}$$

Generally, people always suppose the new system is available, that is,

$$P_0(0, 0) = 1, \quad P_1(0, 0) = 0, \quad P_2(0, 0) = 0.$$

4.2.2 Main Results

Using a common reliability index for a single-unit system with repair delay under discrete time, we introduce the model presented earlier in this chapter into its reliability index according to the reliability index concept of continuous-time repairable systems.

4.2.2.1 System Reliability

Definition 4.4: A random variable X of positive integer is used to describe the failure time of a system or unit. The corresponding distribution sequence is as follows:

$$p_k = P\{X = k\} \quad k = 0, 1, 2, \ldots$$

The probability that the system is always normal (not failed) before k (including k) is

$$R(k) = P\{X > k\} = \sum_{l=k+1}^{\infty} p_l.$$

This is defined as the reliability function or the reliability degree of the system.

Definition 4.5: The mean time to the first failure of system is defined as

$$T_M = E(X) = \sum_{l=k+1}^{\infty} lp_l.$$

Theorem 4.3: The model in Section 4.2.1 is used to establish the following two formulas:

$$R(k) = P_0(k, k)$$

$$T_M = \lambda(0) \sum_{j=1}^{\infty} jP_0(j-1, j-1).$$

Prove: The following formula can be obtained according to definitions of reliability and mean time to the first failure of the system:

$$R(k) = P\{X > k\} = P\{Z(k) = Z(k-1) = \ldots = Z(0) = 0\}.$$

In other words,

$$R(k) = P_0(k, k)$$

$$T_M = E(X) = \sum_{j=1}^{\infty} jP \begin{Bmatrix} Z(0) = \ldots = Z(j-1) = 0 \\ Z(j) = 1 \end{Bmatrix}$$

$$= \sum_{j=1}^{\infty} jP\{Z(0) = \ldots = Z(j-1) = 0\} P\{Z(j) = 1 \mid Z(0) = \ldots = Z(j-1) = 0\}$$

$$= \lambda(0) \sum_{j=1}^{\infty} jP_0(j-1, j-1).$$

4.2.2.2 System Availability

Definition 4.6: The probability that the system is in normal status at the time of k is defined as instantaneous availability:

$$A(k) = P\{Z(k) = 0\}.$$

Theorem 4.4: The model established in Section 4.2.1 is used to obtain the following formula:

$$A(k) = \sum_{j=0}^{k} P_0(k, j)$$

Prove: Making use of the probability analysis method according to the definition of system availability,

$$A(k) = P\{Z(k) = 0\} = P\{Z(k) = 0, Z(k-1) = 1\}$$

$$+ P \left\{ \begin{array}{l} Z(k) = Z(k-1) = 0 \\ Z(k-2) = 1 \end{array} \right\} + \ldots + P\{Z(k) = \ldots = Z(0) = 0\}$$

$$= P_0(k,0) + P_0(k,1) + \ldots + P_0(k,k) = \sum_{j=0}^{k} P_0(k,j)$$

(3) Mean renewal times

Definition 4.7: The mean failure frequency of the system within $(0,k)$ is

$$M(k) = E[N(k)].$$

Theorem 4.5: The model established in Section 4.2.1 is used to obtain the following formula:

$$M(k) = P_0(1,1)M(k-1) + P_2(k,k-1) + \sum_{u=0}^{k-2} P_2(k-1-u,k-2-u)\rho(0)\prod_{s=0}^{u-1}(1-\mu(S))$$

$$+ \sum_{l=1}^{k-2}\sum_{u=0}^{l-1} P_2(l-u,l-u-1)\rho(l-u-1)\prod_{s=0}^{u-1}(1-\mu(S))\mu(u)\big[1+M(k-l-2)\big]$$

Prove:

$M(k) = E[N(k)]$

$\quad = P\{Z(1) = Z(0) = 0\}\, E\{N(k)|Z(1) = Z(0) = 0\}$

$\quad + P\{Z(k) = \ldots = Z(1) = 2, Z(0) = 0\}\, E\{N(k)|Z(k) = \ldots = Z(1) = 2, Z(0) = 0\}$

$\quad + P\{Z(k) = Z(k-1) = \ldots = Z(2) = 1, Z(1) = 2, Z(0) = 0\}$

$\quad \times E\{N(k)|Z(k) = Z(k-1) = \ldots = Z(2) = 1, Z(1) = 2, Z(0) = 0\} + \ldots$

$\quad + P\{Z(k) = 1, Z(k-1) = Z(k-2) = \ldots = Z(1) = 2, Z(0) = 0\}$

$\quad \times E\{N(k)|Z(k) = 1, Z(k-1) = Z(k-2) = \ldots = Z(1) = 2, Z(0) = 0\}$

$$+ \sum_{l=1}^{k-2} \left\{ \begin{array}{l} P\{Z(l+2) = 0, Z(l+1) = 1, Z(l) = \ldots = Z(1) = 2, Z(0) = 0\} \\ \times E\{N(k)|Z(l+2) = 0, Z(l+1) = 1, Z(l) = \ldots = Z(1) = 2, Z(0) = 0\} + \ldots \\ + P\{Z(l+2) = 0, Z(l+1) = \ldots = Z(2) = 1, Z(1) = 2, Z(0) = 0\} \\ \times E\{N(k)|Z(l+2) = 0, Z(l+1) = \ldots = Z(2) = 1, Z(1) = 2, Z(0) = 0\} \end{array} \right\}$$

$$= P_0(1,1)M(k-1)+P_2(k,k-1)+\sum_{u=0}^{k-2}P_2(k-1-u,k-2-u)\rho(0)\prod_{s=0}^{u-1}(1-\mu(s))$$

$$+\sum_{l=1}^{k-2}\sum_{u=0}^{l-1}P_2(l-u,l-u-1)\rho(l-u-1)\prod_{s=0}^{u-1}(1-\mu(s))\mu(u)[1+M(k-l-2)]$$

4.2.3 Numerical Example

Example 1

The failure time, repair time, and delay time of a unit are all subject to geometric distribution; $\lambda(k)$, $\mu(k)$, and $\rho(k)$ correspond to constants λ, μ, and ρ, respectively. Suppose $\lambda = 0.3$, $\mu = 0.8$, and $\rho = 0.2$. In this case, the obtained reliability and availability of the system are as shown in Figure 4.4.

Example 2

When

$$\lambda(k)=\begin{cases} 16\wedge k*\exp(-16)/(k!) & k\le 40 \\ 0.07 & k>40 \end{cases}$$

$$\mu(k)=\begin{cases} 17.5\wedge k*\exp(-17.5)/(k!) & k\le 40 \\ 0.06 & k>40 \end{cases}$$

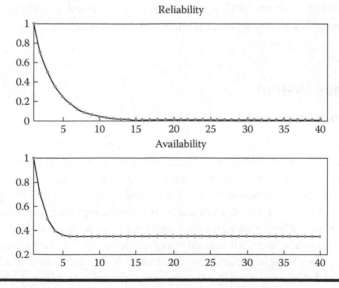

Figure 4.4 Reliability and availability of system when failure time, repair time, and delay time of unit are all subject to geometric distribution.

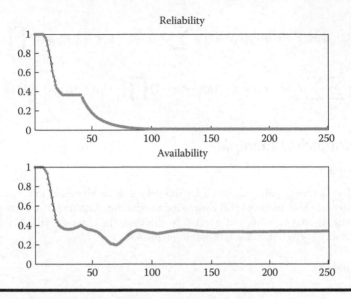

Figure 4.5   **Reliability and availability of special system.**

$$\rho(k) = \left\{ \begin{array}{ll} 13 \wedge k * \exp(-3) / (k!) & k \leq 40 \\ 0.06 & k > 40 \end{array} \right\}$$

The reliability and availability of the system is obtained according to models from the two previous text sections as shown in Figure 4.5.

4.3 Series System

4.3.1 Modeling

Suppose the system is composed of n different units in a series; the failure time X_i and repair time Y_i of unit $i (i = 1, 2, \ldots, n)$ are both subject to geometric distribution; then, its failure rate function and repair rate function are both positive constants, which are expressed as $\lambda_i (k) = \lambda_i$ and $\mu_i(k) = \mu_i$, respectively. Once a unit has failed (one or two), the faulted unit must be repaired immediately after the system is out of order and all other units must be stopped to ensure that no failure appears again, until all faulted units are repaired. Then, n pieces of units immediately enter working state at the same time. When unit i is in normal status at the time of k, record it as $z_i(k) = 0$; otherwise, record it as $z_i(k) = 1$. At the time of k, when all system units are normal, the system is in normal status. In this case, record it as $Z(k) = (0, 0)$; when only units i and j of system are faulted, the system is out of order. At that moment, record it as $Z(k) = (i, j)$; when $i = j$, it means that only unit i or j has failed. In other words, $\{Z(k) = (0, 0)\} \Leftrightarrow \{z_1(k) = z_2(k) = \ldots = z_n(k) = 0\}$, $i, j = 1, 2, \ldots, n, i \leq j$. In order to not lose generality, suppose the system is new at the initial stage; that is, $Z(0) = (0, 0)$. In this chapter, it is supposed that at most only two units can be out of order at the same time. Therefore, the status space of the system can be expressed as follows: $\Omega = \{(i, j)|i, j = 1, 2, \ldots, n,$ and $i \leq j\} \cup \{(0, 0)\}$. The transfer relationship of various kinds of system status is shown in Figure 4.6.

Suppose $P_x(k)$ stands for the probability that system is in a state of $x \in \Omega$ at the time of k; that is,

$$P_{(0, 0)}(k) = P\{Z(k) = (0, 0)\}$$

$$P_{(i, j)}(k) = P\{Z(k) = (i, j)\}, \quad i, j = 1, 2, \ldots, n \text{ and } i \leq j.$$

The transfer model of system status probability can be obtained as follows through probability analysis:

$$P_{(0,0)}(k+1) = P_{(0,0)}(k) \left\{ 1 - \sum_{i=1}^{n} \left[\lambda_t \prod_{\substack{j=1 \\ j \neq i}}^{n} (1-\lambda_t) \right] - \sum_{i=1}^{n-1} \sum_{j=i+1}^{n} \left[\lambda_t \lambda_j \prod_{\substack{l=1 \\ l \neq i,j}}^{n} (1-\lambda_l) \right] \right\}$$

$$+ \sum_{i=1}^{n-1} \sum_{j=i+1}^{n} P_{(i,j)}(k) \mu_i \mu_j + \sum_{i=1}^{n} P_{(i,i)}(k) \mu_i. \tag{4.1}$$

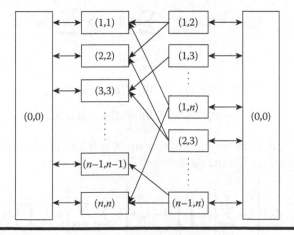

Figure 4.6 Transfer relationship of system status.

$$P_{(i,i)}(k+1) = P_{(0,0)}(k)\lambda_i \prod_{\substack{j=1, \\ j\neq i}}^{n}(1-\lambda_j) + P_{(i,i)}(k)(1-\mu_i)$$

$$+ \left[\sum_{j=1}^{i-1}P_{(j,i)}(k)\mu_j + \sum_{j=i+1}^{n}P_{(i,j)}(k)\mu_j\right](1-\mu_i), \quad i=1,2,\dots,n. \tag{4.2}$$

$$P_{(i,j)}(k+1) = P_{(i,j)}(k)(1-\mu_i)(1-\mu_j) + P_{(0,0)}(k)\lambda_i\lambda_j \prod_{\substack{l=1 \\ l\neq i,j}}^{n}(1-\lambda_l), \quad 1\leq i<j\leq n. \tag{4.3}$$

Next, we will prove these three formulas, respectively.

1. According to the definition, $P_{(0,0)}(k+1)$ stands for the probability that all units of the system are normal at the time of $(k+1)$; that is,

$$P_{(0,0)}(k+1) = P\{Z(k+1) = (0,0)\}.$$

Obviously,

$$\{Z(k+1) = (0,0)\} = B_0 \cup \left(\bigcup_{i=1}^{n}\bigcup_{j=i}^{n}B_{ij}\right).$$

In the formula,

$$B_0 = \{Z(k) = (0,0), Z(k+1) = (0,0)\}$$

$$B_{ij} = \{Z(k) = (i,j), Z(k+1) = (0,0)\}.$$

In addition, B_0 and $B_{ij}(1 \leq i \leq j \leq n)$, are incompatible. Then,

$$P_{(0,0)}(k+1) = PB_0 + \sum_{i=1}^{n}PB_{ii} + \sum_{i=1}^{n-1}\sum_{j=i+1}^{n}PB_{ij}$$

and

$$PB_0 = P\{Z(k) = (0,0), Z(k+1) = (0,0)\}$$

$$= P\{Z(k) = (0,0)\}\, P\{Z(k+1) = (0,0)|Z(k) = (0,0)\}.$$

The fact that, at most, only two units will be out of service at the same time is considered in the following:

$$PB_0 = P_{(0,0)}(k)\left\{1 - \sum_{i=1}^{n}\left[\lambda_t\prod_{\substack{j=1 \\ j\neq i}}^{n}(1-\lambda_t)\right] - \sum_{i=1}^{n-1}\sum_{j=i+1}^{n}\left[\lambda_t\lambda_j\prod_{\substack{l=1 \\ l\neq i,j}}^{n}(1-\lambda_l)\right]\right\}$$

$$PB_{ii} = \Pr\{Z(k) = (i, i), Z(k + 1) = (0, 0)\}$$
$$= P\{Z(k) = (i, i)\}\, P\{Z(k + 1) = (0, 0)|Z(k) = (i, i)\}$$
$$= P_{(i, i)}(k)\mu_i, \quad i = 1, 2, \ldots, n$$
$$PB_{ij} = P\{Z(k) = (i, j), Z(k + 1) = (0, 0)\}$$
$$= P\{Z(k) = (i, i)\}\, P\{Z(k + 1) = (0, 0)|Z(k) = (i, j)\}$$
$$= P_{(i, j)}(k)\mu_i\mu_j, \quad 1 \le i < j \le n.$$

Thus, Formula 4.1 is proven.

2. According to the definition, $P_{(ij)}(k + 1)$ stands for the probability that only unit i of the system has failed at the time of $k + 1$; that is,

$$i = 1, 2, \ldots, n.\ P_{(i, i)}(k + 1) = P\{Z(k + 1) = (i, i)\}.$$

Obviously, $\left\{ Z(k+1) = (i, i) \right\} = \overset{n}{\underset{j=0}{\cup}} C_j$, in which

$$C_0 = \{Z(k + 1) = (i, i), Z(k) = (0, 0)\}$$
$$C_j = \{Z(k + 1) = (i, i), Z(k) = (j, i)\}\ \ j \le i$$
$$C_j = \{Z(k + 1) = (i, i), Z(k) = (i, j)\}\ \ j > i.$$

Also, $C_j(j = 0, 1, 2, \ldots, n)$ are incompatible. Then,

$$P_{(i,i)}(k+1) = \sum_{j=0}^{n} PC_j$$

and

$$PC_0 = P\left\{ Z(k+1) = (i,i), Z(k) = (0,0) \right\}$$
$$= P\left\{ Z(k) = (0,0) \right\} P\left\{ Z(k+1) = (i,i)\,|\,Z(k) = (0,0) \right\}$$
$$= P_{(0,0)}(k)\lambda_i \prod_{j=1,\,j\ne i}^{n} (1 - \lambda_j)$$
$$PC_i = P\left\{ Z(k+1) = (i,i), Z(k) = (i,i) \right\}$$
$$= P\left\{ Z(k) = (i,i) \right\} P\left\{ Z(k+1) = (i,i)\,|\,Z(k) = (i,i) \right\}$$
$$= P_{(i,i)}(k)(1 - \mu_i).$$

When $j < i$,

$$PC_j = P\{Z(k + 1) = (i, i), Z(k) = (j, i)\}$$
$$= P\{Z(k) = (j, i)\}\, P\{Z(k + 1) = (i, i)|Z(k) = (j, i)\}$$
$$= P_{(j, i)}(k)\mu_j(1 - \mu_i).$$

When $j > i$,

$$
\begin{aligned}
PC_j &= P\{Z(k+1) = (i, i), Z(k) = (i, j)\} \\
&= P\{Z(k) = (i, j)\}\, P\{Z(k+1) = (i, i)|Z(k) = (i, j)\} \\
&= P_{(i, j)}(k)\mu_j(1 - \mu_i).
\end{aligned}
$$

Thus, Formula 4.2 is proven.

3. According to the definition, $P_{(i, j)}(k+1)$ stands for the probability that only units i and j of the system have failed at the time of $k + 1$; that is,

$$
i = 1, 2, \ldots, n. \quad P_{(i, j)}(k+1) = P\{Z(k+1) = (i, j)\}.
$$

Obviously, $\{Z(k+1) = (i, j)\} = D_1 \cup D_2$, in which

$$
D_1 = \{Z(k+1) = (i, j), Z(k) = (i, j)\}
$$
$$
D_2 = \{Z(k+1) = (i, j), Z(k) = (0, 0)\}.
$$

Also, D_1 and D_2 are incompatible. Then,

$$
P_{(i, j)}(k+1) = PD_1 + PD_2
$$

and

$$
\begin{aligned}
PD_1 &= P\left\{Z(k+1) = (i, j), Z(k) = (i, j)\right\} \\
&= P\left\{Z(k) = (i, j)\right\} P\left\{Z(k+1) = (i, j) \mid Z(k) = (i, j)\right\} \\
&= P_{(i, j)}(k)(1-\mu_i)(1-\mu_j) \\
PD_2 &= P\left\{Z(k+1) = (i, j), Z(k) = (0,0)\right\} \\
&= P\left\{Z(k) = (0,0)\right\} P\left\{Z(k+1) = (i, j) \mid Z(k) = (0,0)\right\} \\
&= P_{(0,0)}(k)\lambda_i\lambda_j \prod_{\substack{l=1 \\ l \neq i, j}}^{n} (1-\lambda_l)
\end{aligned}
$$

Thus, Formula 4.3 is proven. Generally, people suppose the new system is always available; that is,

$$
P_{(0, 0)}(0) = 1, \quad P_{(i, j)}(0) = 0 \tag{4.4}
$$

Also, when $k \leq t$,

$$P_{(i,j)}(k, t) = 0. \quad i, j = 1, 2, \ldots, n \text{ and } i \leq j. \tag{4.5}$$

4.3.2 Common Reliability Index

A group of common reliability index of series repairable systems composed of n different units under discrete time is derived through the models (4.1 through 4.5) in Section 4.3.1.

4.3.2.1 System Reliability

The reliability function or the reliability of the system is $R(k) = P\{X \geq k\}$. Obviously, $R(0) = 1$.

When $k \geq 1$, the following formula can be obtained:

$$
\begin{aligned}
R(k) &= P\{Z(0) = \ldots = Z(k) = (0, 0)\} \\
&= P\{Z(0) = \ldots = Z(k-1) = (0, 0)\}\, P\{Z(k) = (0, 0) | Z(0) = \ldots = Z(k-1) = (0, 0)\} \\
&= R(k-1)\, P\{Z(k) = (0, 0) | Z(k-1) = (0, 0)\} \\
&= R(k-1)\left(1 - \sum_{i=1}^{n}\left[\lambda_t \prod_{\substack{j=1 \\ j \neq i}}^{n}(1-\lambda_t)\right] - \sum_{i=1}^{n-1}\sum_{j=i+1}^{n}\left[\lambda_t \lambda_j \prod_{\substack{l=1 \\ l \neq i,j}}^{n}(1-\lambda t)\right]\right)
\end{aligned}
$$

Thus,

$$
R(k) = \left\{1 - \sum_{i=1}^{n}\left[\lambda_t \prod_{\substack{j=1 \\ j \neq i}}^{n}(1-\lambda_t)\right] - \sum_{i=1}^{n-1}\sum_{j=i+1}^{n}\left[\lambda_t \lambda_j \prod_{\substack{l=1 \\ l=i,j}}^{n}(1-\lambda t)\right]\right\}^{k}
$$

4.3.2.2 Mean Time to the First Failure of System

Record the mean time to the first failure of system as $T_M = E(X)$.

Then,

$$T_M = E(X) = \sum_{k=0}^{+\infty} k p_k$$

$$= \sum_{k=0}^{+\infty}\left(kP\{Z(0) = \ldots = Z(k) = (0,0), Z(k+1) \neq (0,0)\}\right).$$

In the formula,

$$P\{Z(0) = ... = Z(k) = (0,0), Z(k+1) \neq (0,0)\}$$

$$= \sum_{i=1}^{n} P\{Z(0) = ... = Z(k) = (0,0)\}$$

$$\times P\{Z(k+1) = (i,i) \mid Z(0) = ... = Z(k) = (0,0)\}$$

$$+ \sum_{i=1}^{n-1}\sum_{j=i+1}^{n} \left[\begin{array}{c} P\{Z(0) = ... = Z(k) = (0,0)\} \\ P\{Z(k+1) = (i,j) \mid Z(0) = ... = Z(k) = (0,0)\} \end{array} \right]$$

$$= R(k) \left\{ \sum_{i=1}^{n} \left[\lambda_t \prod_{\substack{j=1 \\ j \neq i}}^{n} (1-\lambda_t) \right] + \sum_{i=1}^{n-1}\sum_{j=i+1}^{n} \left[\lambda_t \lambda_j \prod_{\substack{l=1 \\ l \neq i,j}}^{n} (1-\lambda_l) \right] \right\}$$

Therefore,

$$T_M = \sum_{k=0}^{+\infty} \left\{ kR(k) \left\{ \sum_{i=1}^{n} \left[\lambda_t \prod_{\substack{j=1 \\ j \neq i}}^{n} (1-\lambda_t) \right] + \sum_{i=1}^{n-1}\sum_{j=i+1}^{n} \left[\lambda_t \lambda_j \prod_{\substack{l=1 \\ l \neq i,j}}^{n} (1-\lambda_t) \right] \right\} \right\}.$$

4.3.2.3 System Instantaneous Availability

The instantaneous availability of a system at the time of k is defined as

$$A(k) = P\{Z(k) = (0, 0)\} = P_{(0,\,0)}(k).$$

Obviously,

$$A(k) = P_{(0,\,0)}(k).$$

4.3.3 Numerical Example

Suppose some weapon systems are series repairable systems composed of five differ-ent subsystems under discrete time, as shown in Figure 4.7.

For the failure rate and repair rate of such subsystems, please refer to Table 4.1. The unit of time is days.

Figure 4.7 Series repairable.

Table 4.1 Values of Failure Rate and Repair Rate of Each System Unit

System Unit	Failure Rate	Repair Rate
Unit 1	0.0005	0.05
Unit 2	0.001	0.07
Unit 3	0.0013	0.09
Unit 4	0.0017	0.11
Unit 5	0.002	0.13

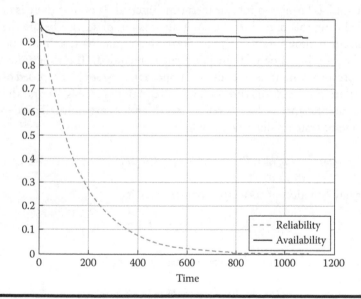

Figure 4.8 Curves of system reliability and availability.

In order to explain the effectiveness of the described model, only two reliability indexes of system (i.e., availability and instantaneous availability) are researched here. The variation of reliability and availability of weapon system in three years is the main focus. The result is shown in Figure 4.8.

From Figure 4.8, it can be seen that system reliability is monotonically decreasing and approaches to zero. In fact, the reliability of the weapon system is almost reduced to zero on the 800th day. It can be interpreted as the weapon system almost cannot be used again after two years of service if it is not repaired after being out of service. The curve of system availability enters service status from the initial moment 1; if it is repaired immediately after it is out of service and the repair method outlined in this chapter is used, then the system availability can be kept above 0.9, which reflects the necessity of repair as well as the quantitative relation between repair and weapon system availability to some extent. Conclusions drawn from models in this chapter are also consistent with actual situations. Therefore, the application of models in this chapter is reasonable to a certain extent.

4.4 Repairable System with Preventive Maintenance Considered

4.4.1 Modeling

Preventive maintenance means that the equipment has to be preventively maintained after it has been used for specified time interval T_0 even if there is no failure appearing during this period. If the equipment is out of service before the expiration of specified time interval T_0, the corrective maintenance should be carried out and the working time of the equipment re-recorded. The time interval T_0 is called the preventive maintenance cycle. Suppose the system is composed of a single unit; its failure time is X; the time of corrective maintenance is Y_1; and the time of preventive maintenance is Y_2. Therefore, X, Y_1, and Y_2 are mutually independent. The function of system failure rate would be

$$\lambda(k) = P\{X = k | X \geq k\} \quad k = 0, 1, 2, \ldots;$$

the repair rate function of corrective maintenance is

$$\mu_1(k) = P\{Y_1 = k | Y_1 \geq k\} \quad k = 0, 1, 2, \ldots;$$

and the repair rate function of preventive maintenance is

$$\mu_2(k) = P\{Y_2 = k | Y_2 \geq k\} \quad k = 0, 1, 2, \ldots;$$

In order to distinguish the different status of the system, the following parameters are defined:

$$\begin{cases} Z(k) = 0 & \textit{System normal at the time k} \\ Z(k) = 1 & \textit{System repaired due to failure at the time k} \quad k = 0, 1, 2, \ldots \\ Z(k) = 2 & \textit{System preventively repaired at the time k} \end{cases}$$

Therefore, make $P_0(k, j)$ stand for the probability that the system has been in 0 status for j unit time at the time of k, $P_1(k, j)$ stand for the probability that the system has been in 1 status for j unit time at the time of k, and $P_2(k, j)$ stand for the probability that system has been in 2 status for j unit time at the time of k with $j = 1$, 2, ..., k, i.e., when $j = 1, 2, ..., k - 1$,

$$
\begin{cases}
P_0(k, j) = P\{Z(k) = ... = Z(k - j) = 0, Z(k - j - 1) \neq 0\} \\
P_1(k, j) = P\{Z(k) = ... = Z(k - j) = 1, Z(k - j - 1) = 0\} \\
P_2(k, j) = P\{Z(k) = ... = Z(k - j) = 2, Z(k - j - 1) = 0\}
\end{cases}
$$

When $j = k$,

$$
\begin{cases}
P_0(k, k) = P\{Z(k) = Z(k - 1) = Z(0) = 0\} \\
P_1(k, k) = P\{Z(k) = Z(k - 1) = Z(0) = 1\} \\
P_2(k, k) = P\{Z(k) = Z(k - 1) = Z(0) = 2\}
\end{cases}
$$

And, when $j > k$, $P_0(k, j) = 0$, $P_1(k, j) = 0$ and $P_2(k, j) = 0$.

The status transfer equation of the system can be obtained as follows through probability analysis:

$$
\begin{cases}
P_0(k + 1, j + 1) = P_0(k, j)(1 - \lambda(j)) & j \leq \min(T_0 - 1, k) \\
P_1(k + 1, j + 1) = P_1(k, j)(1 - \mu_1(j)) & j \leq k \\
P_2(k + 1, j + 1) = P_2(k, j)(1 - \mu_2(j)) & j \leq k
\end{cases}
\tag{4.6}
$$

$$
P_0(k, j) = 0, \quad j > T_0
\tag{4.7}
$$

$$
\begin{cases}
P_0(k + 1, 0) = \sum_{j=0}^{k} (\mu_1(j) P_1(k, j) + \mu_2(j) P_2(k, j)) \\
P_1(k + 1, 0) = \sum_{j=0}^{T_0} \lambda(j) P_0(k, j) \\
P_2(k + 1, 0) = (1 - \lambda(T_0)) P_0(k, T_0)
\end{cases}
\tag{4.8}
$$

Prove: When $j \leq k$, make

$$
B_1 = \{Z(k + 1) = ... = Z(k - j) = 1, Z(k - j - 1) = 0\}
$$
$$
B_2 = \{Z(k) = ... = Z(k - j) = 1, Z(k - j - 1) = 0\}.
$$

Obviously, $B_1 \subset B_2$. Then, $P\{B_1\} = P\{B_2\}\, P\{B_1|B_2\}$. And

$$P\{B_1\} = P_1(k + 1, j + 1)$$
$$P\{B_2\} = P_1(k, j)$$
$$P\{B_1|B_2\} = 1 - \mu_1(j).$$

Thus,

$$P_1(k + 1, j + 1) = P_1(k, j)(1 - \mu_1(j)), \quad j \leq k.$$

Similarly,

$$P_2(k + 1, j + 1) = P_2(k, j)(1 - \mu_2(j)), \quad j \leq k$$
$$P_0(k + 1, j + 1) = P_0(k, j)(1 - \lambda(j)), \quad j \leq \min(T_0 - 1, k).$$

Thus, Formula 4.6 is proven.

When the unit has worked normally and continuously for T_0 time, preventive maintenance is carried out for the system. Thus, the probability that the unit is continuously in 0 status for $j(> T_0)$ unit time is zero. According to the definition of $P_0(k, j)$, $P_0(k, j) = 0$ and $j > T_0$. Thus, Formula 4.7 is proven.

Because

$$P_0(k + 1, 0) = P\{Z(k + 1) = 0, Z(k) \neq 0\}$$
$$= P\{Z(k + 1) = 0, Z(k) = 1\} + P\{Z(k + 1) = 0, Z(k) = 2\},$$

and

$$\{Z(k + 1) = 0, Z(k) = 1\} = \{Z(k + 1) = 0, Z(k) = 1, Z(k - 1) = 0\}$$
$$+ \{Z(k + 1) = 0, Z(k) = Z(k - 1) = 1, Z(k - 2) = 0\} + \dots$$
$$+ \{Z(k + 1) = 0, Z(k) = Z(k - 1) = \dots = Z(0) = 1\}$$

The following formula can be obtained using the method of proof similar to that of Formula 4.6:

$$P\{Z(k+1) = 0, Z(k) = 1\} = \sum_{j=0}^{k} \mu_1(j) P_1(k, j).$$

Similarly,

$$P\{Z(k+1) = 0, Z(k) = 2\} = \sum_{j=0}^{k} \mu_2(j) P_2(k, j).$$

Thus,

$$P_0(k+1,0) = \sum_{j=0}^{k} (\mu_1(j)P_1(k,j) + \mu_2(j)P_2(k,j)).$$

Therefore, the following formula can be obtained:

$$P_1(k+1,0) = \sum_{j=o}^{T_0} \lambda(j)P_0(k,j).$$

According to the definition of preventive maintenance,

$$P_2(k+1, 0) = (1 - \lambda(T_0))\, P_0(k, T_0).$$

which proves Formula 4.8.

4.4.2 System Instantaneous Availability

Generally, suppose the system is new at the initial stage; that is,

$$(P_0(0, 0), P_1(0, 0), P_2(0, 0)) = (1, 0, 0). \tag{4.9}$$

According to the definition, the instantaneous availability of the system is

$$A(k) = P\{Z(k) = 0\}.$$

In the formula,

$$\{Z(k) = 0\} = \{Z(k) = 0, Z(k-1) \neq 0\}$$
$$+ \{Z(k) = Z(k-1) = 0, Z(k-2) \neq 0\} + \ldots.$$
$$+ \{Z(k) = Z(k-1) = \ldots = Z(0) = 0\}.$$

The following formula can be obtained according to models (4.6 through 4.8) presented in this chapter:

$$A(k) = \sum_{j=0}^{k} P_0(k,j) = \sum_{j=0}^{T_0} P_0(k,j) \tag{4.10}$$

Using these models, the mean system availability during $[0, k]$ can be obtained as follows:

$$\bar{A}(T_0) = \frac{1}{k}\sum_{i=0}^{k} A(i) = \frac{1}{k}\sum_{i=0}^{k}\sum_{j=0}^{T_0} P_0(i,j)$$

4.4.3 Numerical Example

Three numerical examples are given to describe models in this section. Suppose both corrective maintenance time and preventive maintenance time are subject to geometric distribution (i.e., the repair rate is positive constant), and these are $\mu_1(k) = 0.01$ and $\mu_2(k) = 0.5$, respectively; failure time of unit is subject to discrete Weibull distribution with parameter (q, β); the scale parameter $q = 0.999$. Three examples are used for research according to the monotonicity (reflected by shape parameter β) of system failure rate; preventive maintenance cycles for these three examples are $T_0 = 3$, $T_0 = 7$, and $T_0 = 11$, respectively.

Example 1

When $\beta = 1$ and the distribution of unit failure time is degraded into geometric distribution, the failure rate $\lambda(k) = 1 - q = 0.001$ is a constant.

Example 2

When $\beta = 1.5 > 1$, the failure rate is a monotone increasing sequence.

Example 3

When $\beta = 0.8 < 1$, the failure rate is a monotone decreasing sequence.

Availability curves of these three examples are presented according to the instantaneous availability model (4.10), as shown in Figures 4.9 through 4.11, respectively.

If the unit still has no failure after being used for T_0 unit time, preventive maintenance should take place. Compare given instantaneous availability curves: the instantaneous availability is usually low when the service time of a system is around integral multiples of unit preventive cycle, which is due to preventive maintenance of a normal unit. The preventive maintenance will make the age of the unit rollback. When the failure rate of a unit is an increasing function, the reduction of service life can decrease the failure rate of the unit, thus improving the steady-state availability. Therefore, the system instantaneous availability can be improved to a certain degree most of the time. Because preventive maintenance will result in an increase of downtime, too frequent preventive maintenance will also decrease availability. From the steady-state availability in Figure 4.10, it can be seen that the case with preventive maintenance cycle = 7 is much better than cases with preventive maintenance cycles = 3 and 11, respectively. Therefore, the system has an optimal preventive maintenance cycle. The proper preventive maintenance cycle is helpful for improving system availability. However, when the unit failure rate is a constant or decreasing function, the preventive maintenance of a normal unit will not improve the unit status but will make a normal unit enter preventive maintenance status, thus reducing availability. In particular, with a reduction in the preventive maintenance cycle or an increase in preventive maintenance frequency, the system availability will be further reduced, as shown in Figures 4.9 and 4.11.

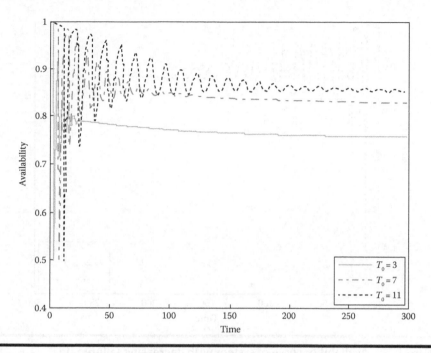

Figure 4.9 **Availability curve of system with constant failure rate.**

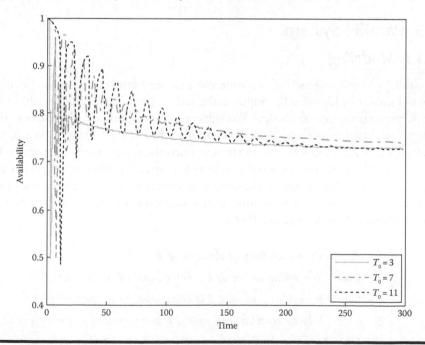

Figure 4.10 **Availability curve of system with increasing failure rate.**

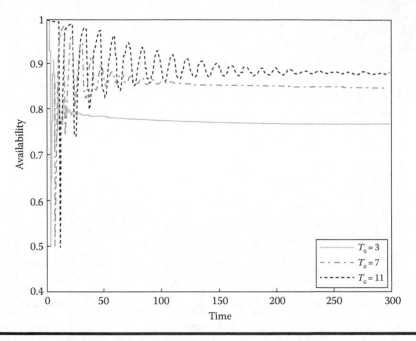

Figure 4.11   Availability curve of system with decreasing failure rate.

4.5 Parallel System

4.5.1 Modeling

A parallel system composed of two different units and one set of repairing equipment is shown in Figure 4.12. Suppose the failure time X_i of unit i is subject to geometric distribution; repair time Y_i is subject to general discrete distribution; the failure rate function and repair rate function of each unit are $\lambda(k)$ and $\mu(k)$, respectively, $k = 0, 1, 2, ...$; these two units are new when the system starts to operate; X_1, X_2, Y_1, and Y_2 are all mutually independent; and after the failed unit is repaired, the distribution of failure time is the same as that of new unit. Therefore, when two units are out of order at the same time, unit 1 should be repaired first, then unit 2. System status $Z(k)$ is defined as follows:

$$
Z(k) = \begin{cases}
0 & \textit{Both units are working at the time of } k \\
1 & \textit{Unit 2 is working and unit 1 is being repaired at the time of } k \\
2 & \textit{Unit 1 is working and unit 2 is being repaired at the time of } k \\
3 & \textit{Unit 1 is being repaired and unit 2 is to be repaired at the time of } k \\
4 & \textit{Unit 2 is being repaired and unit 1 is to be repaired at the time of } k
\end{cases}
$$

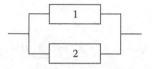

Figure 4.12 Parallel system composed of two units.

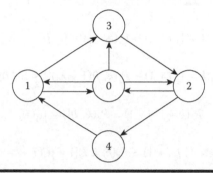

Figure 4.13 Status transfer relationship of discrete system.

The status transfer relationship of a discrete system is shown in Figure 4.13. These units are different from a continuous situation because they may fail at the same time. So, status 0 can directly transfer to status 3.

Therefore, make $P_0(k)$ stand for the probability that system has been in 0 status at the time of k, $P_1(k, j)$ standard for the probability that system has been in 1 status for j unit time at the time of k, $P_2(k, j)$ standard for the probability that system has been in 2 status for j unit time at the time of k, $P_3(k, j)$ stand for the probability that system directly transfers to status 3 from status 0 and has been in status 3 for and only for j unit time at the time of k, $P_3(k, j, l)$ stand for the probability that system has been in 1 status for j unit time at the time of k and transfers to 3 status for l unit time, and $P_4(k, j, l)$ stand for the probability that the system has been in 2 status for j unit time at the time of k and transfers to 4 status for l unit time with $j = 0, 1, \ldots, k - 1$ and $l = 0, 1, \ldots, k - j - 1$. That is,

$$P_0(k) = P\{Z(k) = 0\}$$

$$P_1(k, j) = P\{Z(k) = \ldots = Z(k - j) = 1, Z(k - j - 1) = 0\}$$

$$P_2(k, j) = P\{Z(k) = \ldots = Z(k - j) = 2, Z(k - j - 1) = 0\}$$

$$P_3(k, j) = P\{Z(k) = \ldots = Z(k - j) = 3, Z(k - j - 1) = 0\}$$

$$P_3(k, j, l) = P\{Z(k) = \ldots = Z(k - l) = 3, Z(k - l - 1) = \ldots = Z(k - l - j) = 1, Z(k - l - j - 1) = 0\}$$

$$P_4(k, j, l) = P\{Z(k) = \ldots = Z(k - l) = 4, Z(k - l - 1) = \ldots = Z(k - l - j) = 2, Z(k - l - j - 1) = 0\}$$

Then,

$$P_0(k+1) = P_0(k)(1-\lambda_1)(1-\lambda_2) + \sum_{j=0}^{k-1} P_1(k,j)\mu_1(j)(1-\lambda_2)$$

$$+ \sum_{j=0}^{k-1} P_2(k,j)\mu_2(j)(1-\lambda_1).$$

(4.11)

$$P_1(k+1, j+1) = P_1(k, j)[1 - \mu_1(j)](1 - \lambda_2) \tag{4.12}$$

$$P_2(k+1, j+1) = P_2(k, j)(1 - \lambda_1)[1 - \mu_2(j)] \tag{4.13}$$

$$P_3(k+1, j+1) = P_3(k, j)[1 - \mu_1(j)] \tag{4.14}$$

$$P_3(k+1, j, l+1) = P_3(k, j, l)[1 - \mu_1(j+l+1)] \tag{4.15}$$

$$P_4(k+1, j, l+1) = P_4(k, j, l)[1 - \mu_2(j+l+1)] \tag{4.16}$$

Boundary

$$P_1(k+1,0) = P_0(k)\lambda_1(1-\lambda_2) + \sum_{j=0}^{k-1}\sum_{l=0}^{k-j-1} P_4(k,j,l)\mu_2(j+l+1). \tag{4.17}$$

$$P_2(k+1,0) = P_0(k)(1-\lambda_1)\lambda_2 + \sum_{j=0}^{k-1}\sum_{l=0}^{k-j-1} P_3(k,j,l)\mu_1(j+l+1)$$

$$+ \sum_{j=0}^{k-1} P_3(k,j)\mu_1(j).$$

(4.18)

$$P_3(k, 0) = P_0(k)\lambda_1\lambda_2 \tag{4.19}$$

$$P_3(k+1, j, 0) = P_1(k, j)[1 - \mu_1(j)]\lambda_2 \tag{4.20}$$

$$P_4(k+1, j, 0) = P_1(k, j)\lambda_1[1 - \mu_2(j)] \tag{4.21}$$

Prove:

$$P_0(k+1) = P\{Z(k+1) = 0\}$$
$$= P\{Z(k+1) = Z(k) = 0\} + P\{Z(k+1)$$
$$= 0, Z(k) = 1\} + P\{Z(k+1) = 0, Z(k) = 2\}.$$

In the formula,

$$P\{Z(k + 1) = Z(k) = 0\} = P\{Z(k) = 0\}\, P\{Z(k + 1) = 0|Z(k) = 0\}$$

$$= P_0(k)(1 - \lambda_1)(1 - \lambda_2)$$

$$P\{Z(k+1) = 0, Z(k) = 1\}$$

$$= \sum_{j=0}^{k-1} P\{Z(k+1) = 0, Z(k) = \dots = Z(k-j) = 1, Z(k-j-1) = 0\}$$

$$+ \sum_{j=0}^{k-1} P\{Z(k+1) = 0, Z(k) = \dots = Z(k-j) = 2, Z(k-j-1) = 0\}$$

In the formula,

$$P\{Z(k + 1) = 0,\ Z(k) = \dots = Z(k - j) = 1,\ Z(k - j - 1) = 0\}$$

$$= P\{Z(k) = \dots = Z(k - j) = 1,\ Z(k - j - 1) = 0\}$$

$$\times\ P\{Z(k + 1) = 0|Z(k) = \dots = Z(k - j) = 1,\ Z(k - j - 1) = 0\}$$

$$= P_1(k, j)\, \mu_1(j)\, (1 - \lambda_2)$$

$$P\{Z(k + 1) = 0,\ Z(k) = \dots = Z(k - j) = 2,\ Z(k - j - 1) = 0\}$$

$$= P\{Z(k) = \dots = Z(k - j) = 2,\ Z(k - j - 1) = 0\}$$

$$\times\ P\{Z(k + 1) = 0|Z(k) = \dots = Z(k - j) = 2,\ Z(k - j - 1) = 0\}$$

$$= P_2(k, j)\, \mu_2(j)(1 - \lambda_1)$$

Thus, Formula 4.11 is proven.

$$P_1(k + 1, j + 1) = P\{Z(k + 1) = \dots = Z(k - j) = 1,\ Z(k - j - 1) = 0\}$$

$$= P\{Z(k) = \dots = Z(k - j) = 1,\ Z(k - j - 1) = 0\}$$

$$\times\ P\{Z(k + 1) = 1|Z(k) = \dots = Z(k - j) = 1,\ Z(k - j - 1) = 0\}$$

$$= P_1(k, j)[1 - \mu_1(j)](1 - \lambda_2)$$

Formula 4.12 is proven. Formulas 4.13 through 4.21 are proven in the same way.

4.5.2 System Instantaneous Availability

System instantaneous availability at the time of k is

$$A(k) = P_0(k) + \sum_{j=0}^{k} P_1(k,j) + \sum_{j=0}^{k} P_2(k,j)$$

Prove:

$$A(k) = P\{Z(k) = 0\} + P\{Z(k) = 1\} + P\{Z(k) = 2\}$$

$$P\{Z(k) = 0\} = P_0(k)$$

$$P\{Z(k)=1\} = P\{Z(k)=1, Z(k-1)\neq 1\} + P\{Z(k)=Z(k-1)=1, Z(k-2)\neq 1\} + \dots$$

$$+ P\{Z(k)=Z(k-1)=\dots=Z(0)=1\}$$

$$= \sum_{j=0}^{k} P_1(k,j)$$

$$P\{Z(k)=2\} = P\{Z(k)=2, Z(k-1)\neq 2\} + P\{Z(k)=Z(k-1)=2, Z(k-2)\neq 2\} + \dots$$

$$+ P\{Z(k)=Z(k-1)=\dots=Z(0)=2\}$$

$$= \sum_{j=0}^{k} P_2(k,j)$$

4.6 Summary

In this chapter, detailed research on the modeling method of instantaneous availability of a repairable system under discrete time is carried out; status transfer models for a single-unit repairable system, a single-unit repairable system with repair delay, and a single-unit repairable system with preventive maintenance cycle are considered; and a repairable system with multiple units in series and a repairable system with multiple units in parallel under discrete time are established. Based on all of this, important reliability indexes are obtained, such as system reliability, instantaneous availability, steady-state availability, and mean failure frequency. The modeling method proposed in this chapter is simple, easy to understand, and has clear physical significance. Obtained steady-state or average reliability indexes are consistent with conclusions of continuous-time models, which indicates that the discrete-time model and continuous-time model have no qualitative difference from the view of steady-state or average sense, and that they can mutually interchange for use. The result of instantaneous availability obtained from discrete-time models is intuitive, easy to understand, and very suitable for computer processing, which is quite different from results of continuous-time models. The difference between the two models above on the same system will reflect the possibility whether the discrete-time model can substitute continuous-time model to illustrate the instantaneous availability fluctuation. Therefore, it is vital to discuss the mutual relationship between these two kinds of models and to use the established discrete-time model to research the control of instantaneous availability fluctuation.

Chapter 5

Instantaneous Availability Model of System with Finite Time Restrained and Proof of Its Stability

5.1 Introduction

In practical engineering, the service life, repair time after failure, and the logistic time of unit are all limited. In this chapter, the instantaneous availability model of a system with finite time restrained is established based on Chapter 2. The stability of system instantaneous availability is proven with matrix theory (i.e., the existence of steady-state availability); in addition, the expression of system steady-state availability is obtained.

5.2 General Repairable System

5.2.1 Instantaneous Availability Model

This section covers cases where finite time is restrained. For example, suppose the service life and repair time of a unit are limited, that is, there is $n_1, n_2 \in Z^+$, s.t.,

$$\lambda(n_1) = 1, \mu(n_2) = 1.$$

Theorem 5.1: When arbitrary $k_1 > n_1$ and $k_2 > n_2$,

$$P_0(k, k_1) = 0 \text{ and } P_1(k, k_2) = 0.$$

Prove: When arbitrary $k_1 > n_1$, from Section 4.1, we see that

$$P_0(k, k_1) = P_0(k-1, k_1-1)\left[1 - \lambda(k_1 - 1)\right] = \dots$$

$$= P_0(k - k_1 + n_1, n_1) \prod_{j=n_1}^{k_1-1} \left[1 - \lambda(j)\right] = 0.$$

Similarly, it can be proven that when arbitrary $k_2 > n_2$,

$$P_1(k, k_2) = 0.$$

According to Theorem 5.1, for Section 4.1, the research on $P_0(k, k_1)$ and $P_1(k, k_2)$ may not be done. In this case, $k_1 > n_1$ and $k_2 > n_2$.
Suppose

$$P(k) = (P_0(k), \ P_1(k))^T.$$

In the formula,

$$P_0(k) = \left(P_0(k,0), P_0(k,1), \dots, P_0(k,n_1)\right)^T \in [0,1]^{n_1+1}$$

$$P_1(k) = \left(P_1(k,0), P_1(k,1), \dots, P_1(k,n_2)\right)^T \in [0,1]^{n_2+1}.$$

Then,

$$P(k + 1) = BP(k), \quad k = 0, 1, 2, \dots \tag{5.1}$$

$$P(0) = \left(\delta_{1,n_1+n_2+1}\right)^T. \tag{5.2}$$

The system instantaneous availability is as follows:

$$A(k) = \delta_{n_1+1, n_2+1} P(k). \tag{5.3}$$

In the formula,

$$B = \begin{bmatrix} 0 & 0 & \cdots & 0 & 0 & \mu(0) & \mu(1) & \cdots & \mu(n_2-1) & 1 \\ 1-\lambda(0) & 0 & \cdots & 0 & 0 & 0 & 0 & \cdots & 0 & 0 \\ 0 & 1-\lambda(1) & \cdots & 0 & 0 & 0 & 0 & \cdots & 0 & 0 \\ \cdots & \cdots & \cdots & \cdots & \cdots & \cdots & \cdots & \cdots & \cdots & \cdots \\ 0 & 0 & \cdots & 1-\lambda(n_1-1) & 0 & 0 & 0 & \cdots & 0 & 0 \\ \lambda(0) & \lambda(1) & \cdots & \lambda(n_1-1) & 1 & 0 & 0 & \cdots & 0 & 0 \\ 0 & 0 & \cdots & 0 & 0 & 1-\mu(0) & 0 & \cdots & 0 & 0 \\ 0 & 0 & \cdots & 0 & 0 & 0 & 1-\mu(1) & \cdots & 0 & 0 \\ \cdots & \cdots & \cdots & \cdots & \cdots & \cdots & \cdots & \cdots & \cdots & \cdots \\ 0 & 0 & \cdots & 0 & 0 & 0 & 0 & \cdots & 1-\mu(n_2-1) & 0 \end{bmatrix}$$

$$\delta_{m,n} = \left(\overbrace{1,...,1}^{m}, \overbrace{0,...,0}^{n} \right).$$

5.2.2 Stability Proving of Instantaneous Availability

According to features of matrix B and initial value $P(0)$, the following conclusions are set up.

Theorem 5.2: The status of systems (5.1 and 5.2) only exists in finite set C, in which,

$$C = \left\{ Z = (z_1, z_2,...,z_{m_1+n_2+2})^T \mid \delta_{n_1+n_2+2,0} Z = 1, 0 \le z_i \le 1, i = 1,2,...,n_1+n_2+2 \right\}.$$

Prove: $\forall k \in Z^+$

$$\delta_{n_1+n_2+2,0} P(k+1) = \delta_{n_1+n_2+2,0} BP(k).$$

For matrix B, obviously,

$$\delta_{n_1+n_2+2,0} B = \delta_{n_1+n_2+2,0}.$$

Thus,

$$\delta_{n_1+n_2+2,0} P(k+1) = \delta_{n_1+n_2+2,0} P(k) = ... = \delta_{n_1+n_2+2,0} P(0) = 1.$$

Then, for arbitrary nonnegative integer k, the status $P(k) = \left(p_1, p_2, ..., p_{n_1+n_2+2} \right)^T$ satisfies

$$\delta_{n_1+n_2+2,0} P(k) = \sum_{i=1}^{n_1+n_2+2} p_i = 1. \tag{5.4}$$

According to definitions of failure rate and repair rate,

$$0 \le \lambda(k) \le 1, \quad k = 0, 1, 2, ..., n_1$$

$$0 \le \mu(k) \le 1, \quad k = 0, 1, 2, ..., n_2.$$

Then, arbitrary element in matrix B satisfies

$$b_{i,j} \ge 0, \quad i, j = 1, 2, ..., n_1 + n_2 + 2.$$

Therefore,

$$p_i \ge 0, \quad p_i \le 1 \quad i = 1, 2, ..., n_1 + n_2 + 2.$$

The theorem is proven. For systems (5.1 and 5.2), the following formula can be obtained:

$$P(k) = B^k \left(\delta_{1, n_1+n_2+1} \right)^T \quad k = 0, 1, 2, ... \tag{5.5}$$

In other words, it is the status solution of systems (5.1 and 5.2).

Theorem 5.3: All characteristic roots of matrix B are in a unit circle, that is, $\rho(B) \le 1$.

Prove: $B - \gamma I$

$$=
\begin{bmatrix}
-\gamma & 0 & \cdots & 0 & 0 & \mu(0) & \mu(1) & \cdots & \mu(n_2-1) & 1 \\
1-\lambda(0) & -\gamma & \cdots & 0 & 0 & 0 & 0 & \cdots & 0 & 0 \\
0 & 1-\lambda(1) & \cdots & 0 & 0 & 0 & 0 & \cdots & 0 & 0 \\
\cdots & \cdots & \cdots & \cdots & \cdots & \cdots & \cdots & \cdots & \cdots & \cdots \\
0 & 0 & \cdots & 1-\lambda(n_1-1) & -\gamma & 0 & 0 & \cdots & 0 & 0 \\
\lambda(0) & \lambda(1) & \cdots & \lambda(n_1-1) & 1 & -\gamma & 0 & \cdots & 0 & 0 \\
0 & 0 & \cdots & 0 & 0 & 1-\mu(0) & -\gamma & \cdots & 0 & 0 \\
0 & 0 & \cdots & 0 & 0 & 0 & 1-\mu(1) & \cdots & 0 & 0 \\
\cdots & \cdots & \cdots & \cdots & \cdots & \cdots & \cdots & \cdots & \cdots & \cdots \\
0 & 0 & \cdots & 0 & 0 & 0 & 0 & \cdots & 1-\mu(n_2-1) & -\gamma
\end{bmatrix},
$$

when $\forall \gamma \in \{x \mid |x| > 1\}$, $\gamma I - B$ is strictly a diagonally dominant matrix [83]. Therefore, $\gamma I - B$ is nonsingular; that is, $\rho(B) \leq 1$.

Theorem 5.4: Rank $(I - B) = n_1 + n_2 + 1$.

Prove: Matrix

$$
B - I =
\begin{bmatrix}
-1 & 0 & \cdots & 0 & 0 & \mu(0) & \mu(1) & \cdots & \mu(n_2-1) & 1 \\
1-\lambda(0) & -1 & \cdots & 0 & 0 & 0 & 0 & \cdots & 0 & 0 \\
0 & 1-\lambda(1) & \cdots & 0 & 0 & 0 & 0 & \cdots & 0 & 0 \\
\cdots & \cdots & \cdots & \cdots & \cdots & \cdots & \cdots & \cdots & \cdots & \cdots \\
0 & 0 & \cdots & 1-\lambda(n_1-1) & -1 & 0 & 0 & \cdots & 0 & 0 \\
\lambda(0) & \lambda(1) & \cdots & \lambda(n_1-1) & 1 & -1 & 0 & \cdots & 0 & 0 \\
0 & 0 & \cdots & 0 & 0 & 1-\mu(0) & -1 & \cdots & 0 & 0 \\
0 & 0 & \cdots & 0 & 0 & 0 & 1-\mu(1) & \cdots & 0 & 0 \\
\cdots & \cdots & \cdots & \cdots & \cdots & \cdots & \cdots & \cdots & \cdots & \cdots \\
0 & 0 & \cdots & 0 & 0 & 0 & 0 & \cdots & 1-\mu(n_2-1) & -1
\end{bmatrix}.
$$

Then, the sum of elements in all columns of matrix $B - I$ equals 0. Therefore, rank $(B - I) \leq n_1 + n_2 + 1$. In addition, for matrix $B - I$, the principal minor of order $n_1 + n_2 + 1$ is $(B - I)_{(1,\,1)}$:

$$
=
\begin{bmatrix}
-1 & 0 & \cdots & 0 & 0 & 0 & 0 & \cdots & 0 & 0 \\
1-\lambda(1) & -1 & \cdots & 0 & 0 & 0 & 0 & \cdots & 0 & 0 \\
0 & 1-\lambda(2) & \cdots & 0 & 0 & 0 & 0 & \cdots & 0 & 0 \\
\cdots & \cdots & \cdots & \cdots & \cdots & \cdots & \cdots & \cdots & \cdots & \cdots \\
0 & 0 & \cdots & 1-\lambda(n_1-1) & -1 & 0 & 0 & \cdots & 0 & 0 \\
\lambda(1) & \lambda(2) & \cdots & \lambda(n_1-1) & 1 & -1 & 0 & \cdots & 0 & 0 \\
0 & 0 & \cdots & 0 & 0 & 1-\mu(0) & -1 & \cdots & 0 & 0 \\
0 & 0 & \cdots & 0 & 0 & 0 & 1-\mu(1) & \cdots & 0 & 0 \\
\cdots & \cdots & \cdots & \cdots & \cdots & \cdots & \cdots & \cdots & \cdots & \cdots \\
0 & 0 & \cdots & 0 & 0 & 0 & 0 & \cdots & 1-\mu(n_2-1) & -1
\end{bmatrix}.
$$

Obviously, $(B - I)_{(1,1)}$ is nondegenerate. There is rank $(B - I) \geq n_1 + n_2 + 1$. Therefore, rank $(I - B) = n_1 + n_2 + 1$.

Theorem 5.5: Matrix B has the characteristic root $\gamma = 1$, which is a simple root. If $X_0 = (x_1, x_2, \ldots, x_{2n+2})^T$ stands for the feature vector of $\gamma = 1$ on C, the value of X_0 is unique and satisfies

$$x_1 = x_{n_1+2} = \frac{1}{D}$$

$$x_i = \frac{1}{D} \prod_{j=0}^{i-2} \left[1 - \lambda(j)\right], \quad i = 2, 3, \ldots, n_1 + 1$$

$$x_{n_1+i+1} = \frac{1}{D} \prod_{j=0}^{i-2} \left[1 - \mu(j)\right], \quad i = 2, 3, \ldots, n_2 + 1.$$

In the formula,

$$D = 2 + \sum_{i=2}^{n_1+1} \left\{ \prod_{j=0}^{i-2} \left[1 - \lambda(j)\right] \right\} + \sum_{i=2}^{n_2+1} \left\{ \prod_{j=0}^{i-2} \left[1 - \mu(j)\right] \right\}.$$

Prove: According to Theorem 5.4, $\gamma = 1$ is the simple characteristic root of matrix B, which is equivalent to

$$\left. \frac{d|B - \gamma I|}{d\gamma} \right|_{\gamma=1} \neq 0. \tag{5.6}$$

Then, Formula 5.6 is proven:

$$\left. \frac{d|B - \gamma I|}{d\gamma} \right|_{\gamma=1}$$

$$= \begin{bmatrix}
-1 & 0 & \cdots & 0 & 0 & \mu(0) & \mu(1) & \cdots & \mu(n_2-1) & 1 \\
0 & -1 & \cdots & 0 & 0 & 0 & 0 & \cdots & 0 & 0 \\
0 & 1-\lambda(1) & \cdots & 0 & 0 & 0 & 0 & \cdots & 0 & 0 \\
\cdots & \cdots & \cdots & \cdots & \cdots & \cdots & \cdots & \cdots & \cdots & \cdots \\
0 & 0 & \cdots & 1-\lambda(n_1-1) & -1 & 0 & 0 & \cdots & 0 & 0 \\
0 & \lambda(1) & \cdots & \lambda(n_1-1) & 1 & -1 & 0 & \cdots & 0 & 0 \\
0 & 0 & \cdots & 0 & 0 & 1-\mu(0) & -1 & \cdots & 0 & 0 \\
0 & 0 & \cdots & 0 & 0 & 0 & 1-\mu(1) & \cdots & 0 & 0 \\
\cdots & \cdots & \cdots & \cdots & \cdots & \cdots & \cdots & \cdots & \cdots & \cdots \\
0 & 0 & \cdots & 0 & 0 & 0 & 0 & \cdots & 1-\mu(n_2-1) & -1
\end{bmatrix}$$

$$+\begin{bmatrix}
-1 & 0 & \cdots & 0 & 0 & \mu(0) & \mu(1) & \cdots & \mu(n_2-1) & 1 \\
1-\lambda(0) & -1 & \cdots & 0 & 0 & 0 & 0 & \cdots & 0 & 0 \\
0 & 0 & \cdots & 0 & 0 & 0 & 0 & \cdots & 0 & 0 \\
\cdots & \cdots & \cdots & \cdots & \cdots & \cdots & \cdots & \cdots & \cdots & \cdots \\
0 & 0 & \cdots & 1-\lambda(n_1-1) & -1 & 0 & 0 & \cdots & 0 & 0 \\
\lambda(0) & 0 & \cdots & \lambda(n_1-1) & 1 & -1 & 0 & \cdots & 0 & 0 \\
0 & 0 & \cdots & 0 & 0 & 1-\mu(0) & -1 & \cdots & 0 & 0 \\
0 & 0 & \cdots & 0 & 0 & 0 & 1-\mu(1) & \cdots & 0 & 0 \\
\cdots & \cdots & \cdots & \cdots & \cdots & \cdots & \cdots & \cdots & \cdots & \cdots \\
0 & 0 & \cdots & 0 & 0 & 0 & 0 & \cdots & 1-\mu(n_2-1) & -1
\end{bmatrix}+\cdots$$

$$+\begin{bmatrix}
-1 & 0 & \cdots & 0 & 0 & \mu(0) & \mu(1) & \cdots & 0 & 1 \\
1-\lambda(0) & -1 & \cdots & 0 & 0 & 0 & 0 & \cdots & 0 & 0 \\
0 & 1-\lambda(1) & \cdots & 0 & 0 & 0 & 0 & \cdots & 0 & 0 \\
\cdots & \cdots & \cdots & \cdots & \cdots & \cdots & \cdots & \cdots & \cdots & \cdots \\
0 & 0 & \cdots & 1-\lambda(n_1-1) & -1 & 0 & 0 & \cdots & 0 & 0 \\
\lambda(0) & \lambda(1) & \cdots & \lambda(n_1-1) & 1 & -1 & 0 & \cdots & 0 & 0 \\
0 & 0 & \cdots & 0 & 0 & 1-\mu(0) & -1 & \cdots & 0 & 0 \\
\cdots & \cdots & \cdots & \cdots & \cdots & \cdots & \cdots & \cdots & \cdots & \cdots \\
0 & 0 & 0 & 0 & 0 & 0 & 0 & \cdots & -1 & 0 \\
0 & 0 & \cdots & 0 & 0 & 0 & 0 & \cdots & 0 & -1
\end{bmatrix}$$

$$+\begin{bmatrix}
-1 & 0 & \cdots & 0 & 0 & \mu(0) & \mu(1) & \cdots & \mu(n_2-1) & 0 \\
1-\lambda(0) & -1 & \cdots & 0 & 0 & 0 & 0 & \cdots & 0 & 0 \\
0 & 1-\lambda(1) & \cdots & 0 & 0 & 0 & 0 & \cdots & 0 & 0 \\
\cdots & \cdots & \cdots & \cdots & \cdots & \cdots & \cdots & \cdots & \cdots & \cdots \\
0 & 0 & \cdots & 1-\lambda(n_1-1) & -1 & 0 & 0 & \cdots & 0 & 0 \\
\lambda(0) & \lambda(1) & \cdots & \lambda(n_1-1) & 1 & -1 & 0 & \cdots & 0 & 0 \\
0 & 0 & \cdots & 0 & 0 & 1-\mu(0) & -1 & \cdots & 0 & 0 \\
0 & 0 & \cdots & 0 & 0 & 0 & 1-\mu(1) & \cdots & 0 & 0 \\
\cdots & \cdots & \cdots & \cdots & \cdots & \cdots & \cdots & \cdots & \cdots & \cdots \\
0 & 0 & \cdots & 0 & 0 & 0 & 0 & \cdots & 1-\mu(n_2-1) & -1
\end{bmatrix}$$

$$= (-1)^{n_1+n_2+1}\left\{\begin{array}{l}
-2+\left[\lambda(0)-1\right]+\ldots+\left[\lambda(n_1-1)-1\right]+\left[\mu(0)-1\right]+\ldots \\
+\left[\mu(n_2-1)-1\right]
\end{array}\right\} \neq 0.$$

Thus, $\gamma = 1$ is the simple characteristic root of matrix B.

$X_0 B$ stands for the feature vector of $\gamma = 1$ on C. Obviously, $(B-I) X_0 = 0$; that is,

$$
\begin{bmatrix}
-1 & 0 & \cdots & 0 & 0 & \mu(0) & \mu(1) & \cdots & \mu(n_2-1) & 1 \\
1-\lambda(0) & -1 & \cdots & 0 & 0 & 0 & 0 & \cdots & 0 & 0 \\
0 & 1-\lambda(1) & \cdots & 0 & 0 & 0 & 0 & \cdots & 0 & 0 \\
\cdots & \cdots & \cdots & \cdots & \cdots & \cdots & \cdots & \cdots & \cdots & \cdots \\
0 & 0 & \cdots & 1-\lambda(n_1-1) & -1 & 0 & 0 & \cdots & 0 & 0 \\
\lambda(0) & \lambda(1) & \cdots & \lambda(n_1-1) & 1 & -1 & 0 & \cdots & 0 & 0 \\
0 & 0 & \cdots & 0 & 0 & 1-\mu(0) & -1 & \cdots & 0 & 0 \\
0 & 0 & \cdots & 0 & 0 & 0 & 1-\mu(1) & \cdots & 0 & 0 \\
\cdots & \cdots & \cdots & \cdots & \cdots & \cdots & \cdots & \cdots & \cdots & \cdots \\
0 & 0 & \cdots & 0 & 0 & 0 & 0 & \cdots & 1-\mu(n_2-1) & -1
\end{bmatrix}
$$

$\times X_0 = 0.$

According to Theorem 5.4, the rank of matrix $I-B$ is $n_1 + n_2 + 1$. Then,

$$
\left\{
\begin{aligned}
& x_2 = \left[1-\lambda(0)\right]x_1 \\
& x_3 = \left[1-\lambda(0)\right]\left[1-\lambda(1)\right]x_1 \\
& \cdots \\
& x_{n_1+1} = \prod_{i=0}^{n_1-1}\left[1-\lambda(i)\right]x_1 \\
& x_{n_1+3} = \left[1-\mu(0)\right]x_{n_1+2} \\
& x_{n_1+4} = \left[1-\mu(0)\right]\left[1-\mu(1)\right]x_{n_1+2} \\
& \cdots \\
& x_{n_1+n_2+2} = \prod_{i=0}^{n_1-1}\left[1-\mu(i)\right]x_{n_1+2} \\
& \sum_{i=1}^{n_2+1}\left[x_{n_1+i+1}\mu(i-1)\right]- x_1 = 0
\end{aligned}
\right.
$$

that is,

$$x_i = \prod_{j=0}^{i-2}\left[1-\lambda(j)\right]x_1, \quad i=2,3,\ldots, n_2+1$$

$$x_{m+i+1} = \prod_{j=0}^{i-2}\left[1-\mu(j)\right]x_{m+2}, \quad i=2,3,\ldots, n_2+1$$

$$\sum_{i=1}^{n_2+1}\left[x_{m+i+1}\mu(i-1)\right]-x_1 = 0.$$

Suppose

$$D_1 = \mu(0)+\sum_{i=2}^{n_2+1}\left\{\prod_{j=0}^{i-2}\left[1-\mu(j)\right]\mu(i-1)\right\}.$$

Then,

$$x_1 = D_1 x_{m+2}.$$

The following conclusion is easily proved:

$$D_1 = 1.$$

Then,

$$x_1 = x_{m+2}.$$

Because $X_0 \in C$, that is,

$$\left(1+\sum_{i=2}^{n_1+1}\left\{\prod_{j=0}^{i-2}\left[1-\lambda(j)\right]\right\}\right)x_1 + \left(1+\sum_{i=2}^{n_2+1}\left\{\prod_{j=0}^{i-2}\left[1-\mu(j)\right]\right\}\right)x_1 = 1.$$

Suppose

$$D = 2+\sum_{i=2}^{n_1+1}\left\{\prod_{j=0}^{i-2}\left[1-\lambda(j)\right]\right\}+\sum_{i=2}^{n_2+1}\left\{\prod_{j=0}^{i-2}\left[1-\mu(j)\right]\right\}.$$

Then,

$$x_1 = x_{m+2} = \frac{1}{D}x_i = \frac{1}{D}\prod_{j=0}^{i-2}\left[1-\lambda(j)\right], \quad i=2,3,\ldots, n_1+1$$

$$x_{m_1+i+1} = \frac{1}{D} \prod_{j=0}^{i-2} \left[1 - \mu(j) \right], \quad i = 2, 3, \ldots, n_2 + 1.$$

Theorem 5.6: The characteristic root of matrix B that corresponds to $\rho(B) = 1$ can only be $\gamma = 1$.

Prove: Suppose that the characteristic root of matrix B that corresponds to $\rho(B) = 1$ is $\gamma = e^{bw}$, in which, $w^2 = -1$ and $0 \leq b < 2\pi$. Suppose the feature vector corresponding to characteristic root $\gamma = e^{bw}$ of matrix B is $\zeta = (x_1, x_2, \ldots, x_{m+n_2+2})^T \neq 0$. Then,

$$(B - \gamma I)\zeta = 0.$$

In other words,

$$
\begin{cases}
-\gamma x_1 + \sum_{i=1}^{n_2+1} \left[x_{m_1+i+1} \mu(i-1) \right] = 0 \\[2mm]
\left[1 - \lambda(0) \right] x_1 - \gamma x_2 = 0 \\[1mm]
\left[1 - \lambda(1) \right] x_2 - \gamma x_3 = 0 \\
\cdots \\
\left[1 - \lambda(n_1 - 1) \right] x_{n_1} - \gamma x_{m_1+1} = 0 \\[2mm]
\sum_{i=1}^{n_1+1} \left[x_i \lambda(i-1) \right] - \gamma x_{m_1+2} = 0 \\[2mm]
\left[1 - \mu(0) \right] x_{m_1+2} - \gamma x_{m_1+3} = 0 \\[1mm]
\left[1 - \mu(1) \right] x_{m_1+3} - \gamma x_{m_1+4} = 0 \\
\cdots \\
\left[1 - \mu(n_2 - 1) \right] x_{m_1+n_2+1} - \gamma x_{m_1+n_2+2} = 0.
\end{cases}
$$

Then,

$$x_i = \prod_{j=0}^{i-2} \left[\frac{1 - \lambda(j)}{\gamma} \right] x_1, \quad i = 2, 3, \ldots, n_1 + 1$$

$$x_{m_1+i+1} = \prod_{j=0}^{i-2} \left[\frac{1 - \mu(j)}{\gamma} \right] x_{m_1+2}, \quad i = 2, 3, \ldots, n_2 + 1$$

$$\sum_{i=1}^{n_2+1}\left[x_{m+i+1}\frac{\mu(i-1)}{\gamma}\right]-x_1=0$$

$$\sum_{i=1}^{n_1+1}\left[x_i\frac{\lambda(i-1)}{\gamma}\right]-x_{m+2}=0.$$

Suppose

$$D_1=\lambda(0)e^{-bw}+\sum_{i=2}^{n_1+1}\left\{\prod_{j=0}^{i-2}\left[1-\lambda(j)\right]\lambda(i-1)e^{-biw}\right\}$$

$$D_2=\mu(0)e^{-bw}+\sum_{i=2}^{n_2+1}\left\{\prod_{j=0}^{i-2}\left[1-\mu(j)\right]\mu(i-1)e^{-biw}\right\}.$$

Then,

$$x_1=D_2x_{m+2}$$

$$x_{m+2}=D_1x_1.$$

Thus,

$$(D_2\,D_1-1)x_1=0.$$

Obviously, $x_1\neq0$. Otherwise, $x_i=0$ and $i=1,2,\ldots,n_1+n_2+2$, which is contradictory to $\zeta=\left(x_1,x_2,\ldots,x_{2n+2}\right)^T\neq0$.

Then,

$$D_2\,D_1=1.$$

Because

$$|D_2D_1|=|D_2\|D_1|,$$

in the formula,

$$|D_1|=\left|\lambda(0)e^{-bw}+\sum_{i=2}^{n_1+1}\left\{\prod_{j=0}^{i-2}\left[1-\lambda(j)\right]\lambda(i-1)e^{-biw}\right\}\right|$$

$$\leq \lambda(0)\left|e^{-bw}\right| + \sum_{i=2}^{n_1+1}\left\{\prod_{j=0}^{i-2}\left[1-\lambda(j)\right]\lambda(i-1)\left|e^{-biw}\right|\right\}$$

$$= \lambda(0) + \sum_{i=2}^{n_1+1}\left\{\prod_{j=0}^{i-2}\left[1-\lambda(j)\right]\lambda(i-1)\right\} = 1$$

$$|D_2| = \left|\mu(0)e^{-bw} + \sum_{i=2}^{n_2+1}\left\{\prod_{j=0}^{i-2}\left[1-\mu(j)\right]\mu(i-1)e^{-biw}\right\}\right|$$

$$\leq \mu(0)\left|e^{-bw}\right| + \sum_{i=2}^{n_2+1}\left\{\prod_{j=0}^{i-2}\left[1-\mu(j)\right]\mu(i-1)\left|e^{-biw}\right|\right\}$$

$$= \mu(0) + \sum_{i=2}^{n_2+1}\left\{\prod_{j=0}^{i-2}\left[1-\mu(j)\right]\mu(i-1)\right\} = 1.$$

According to Reference [84], the necessary and sufficient conditions for establishment of an equal sign in this formula are as follows:

1. All vectors expressed by $\lambda(0)e^{-bw}$ and $\prod_{j=0}^{i-2}\left[1-\lambda(j)\right]\lambda(i-1)e^{-biw}$ ($i = 2, 3, \ldots,$

   $n_1 + 1$) are collinear in the same direction.

2. All vectors expressed by $\mu(0)e^{-bw}$ and $\prod_{j=0}^{i-2}\left[1-\mu(j)\right]\mu(i-1)e^{-biw}$ ($i = 2, 3, \ldots,$

   $n_2 + 1$) are collinear in the same direction.

According to the value range of $0 \leq b < 2\pi$, $b = 0$ can be obtained. Thus, the characteristic root of matrix B corresponding to $\rho(B) = 1$ can only be $\gamma = 1$.

The following conclusions can be drawn in combining Theorems 5.3–5.6.

Theorem 5.7: The status solution (5.5) of systems (5.1 through 5.3) is stable, and the unique equilibrium point is X_0 (i.e., feature vector of characteristic root $\gamma = 1$ of matrix B on C).

Prove: According to Theorem 5.8.6 in Reference [85], the status solution (5.5) of systems (5.1 through 5.3) is stable. Suppose the equilibrium point is X_1, $X_1 \in C$ and $X_1 = BX_1$. Then, X_1 is the feature vector of characteristic root $\gamma = 1$ of matrix B on C. Then, $X_1 = X_0$ can be obtained as per Theorem 5.5.

Theorem 5.8: System instantaneous availability (5.3) is stable, and the equilibrium point is unique; that is, the steady-state availability of the system is unique with a value as follows:

$$A = \sum_{i=1}^{n_1+1} x_i = \frac{1}{D}\left(1 + \sum_{i=2}^{n_1+1}\left\{\prod_{j=0}^{i-2}[1-\lambda(j)]\right\}\right). \tag{5.7}$$

In the formula,

$$D = 2 + \sum_{i=2}^{n_1+1}\left\{\prod_{j=0}^{i-2}[1-\lambda(j)]\right\} + \sum_{i=2}^{n_2+1}\left\{\prod_{j=0}^{i-2}[1-\mu(j)]\right\}.$$

Prove: According to Theorem 5.7, the following formula can be obtained:

$$X_0 = \lim_{k\to\infty} P(k).$$

According to Theorem 5.5, the limit

$$A = \lim_{k\to\infty} A(k) = \lim_{k\to\infty} \delta_{n_1+1,n_2+1}P(k) = \delta_{n_1+1,n_2+1}X_0 = \sum_{i=1}^{n_1+1} x_i$$

$$= \frac{1}{D}\left(1 + \sum_{i=2}^{n_1+1}\left\{\prod_{j=0}^{i-2}[1-\lambda(j)]\right\}\right)$$

exists (i.e., steady-state availability of system).

Obviously,

$$A = \frac{MTBF}{MTBF + MTTR}.$$

In other words, the steady-state availability and inherent availability of a system are numerically equal. Unless there is a special situation, these two factors will not be differentiated in this chapter.

5.3 Repairable System with Repair Delay

5.3.1 System Instantaneous Availability Model

In this section, a case with finite time restrained is researched. Suppose the service life and repair time of a unit are limited; that is, there is n_1, n_2, $n_3 \in Z^+$, s.t.

$$\lambda(n_1) = 1, \quad \mu(n_2) = 1, \quad \rho(n_3) = 1.$$

Theorem 5.9: When arbitrary $k_1 > n_1$, $k_2 > n_2$, and $k_3 > n_3$, $P_0(k, k_1)=0$, $P_1(k, k_2)=0$ and $P_2(k, k_3)=0$.

Prove: When arbitrary $k_1 > n_1$, the following is true:

$$P_0\left(k, k_1\right) = P_0\left(k-1, k_1-1\right)\left[1-\lambda\left(k_1-1\right)\right] = \dots$$

$$= P_0\left(k-k_1+n_1, n_1\right)\prod_{j=n_1}^{k_1-1}\left[1-\lambda(j)\right] = 0.$$

Similarly, it can be proven that when arbitrary $k_2 > n_2$ and $k_3 > n_3$,

$$P_1(k, k_2)=0 \text{ and } P_2(k, k_3)=0.$$

According to Theorem 5.9, for Section 4.2, the research on $P_0(k, k_1)$, $P_1(k, k_2)$, and $P_2(k, k_3)$ may not be done. In this case, $k_1 > n_1$, $k_2 > n_2$ and $k_3 > n_3$. Therefore, suppose

$$P(k) = (P_0(k), P_1(k), \dots, P_2(k))^T.$$

In the formula,

$$P_0\left(k\right) = \left(P_0\left(k,0\right), P_0\left(k,1\right), \dots, P_0\left(k, n_1\right)\right)^T \in [0,1]^{n_1+1}$$

$$P_1\left(k\right) = \left(P_1\left(k,0\right), P_1\left(k,1\right), \dots, P_1\left(k, n_2\right)\right)^T \in [0,1]^{n_2+1}$$

$$P_2\left(k\right) = \left(P_2\left(k,0\right), P_2\left(k,1\right), \dots, P_2\left(k, n_3\right)\right)^T \in [0,1]^{n_3+1}.$$

Then,

$$P(k+1) = BP(k), \quad k = 0, 1, 2, \dots \tag{5.8}$$

$$P(0) = \left(\delta_{1, n_1+n_2+n_3+2}\right)^T. \tag{5.9}$$

System instantaneous availability is as follows:

$$A(k) = \delta_{n_1+1, n_2+n_3+2} P(k). \tag{5.10}$$

In the formula,

$$
B = \begin{bmatrix}
0 & 0 & \cdots & 0 & 0 & \mu(0) & \mu(1) & \cdots \\
1-\lambda(0) & 0 & \cdots & 0 & 0 & 0 & 0 & \cdots \\
0 & 1-\lambda(1) & \cdots & 0 & 0 & 0 & 0 & \cdots \\
\cdots & \cdots & \cdots & \cdots & \cdots & \cdots & \cdots & \cdots \\
0 & 0 & \cdots & 1-\lambda(n-1_1) & 0 & 0 & 0 & \cdots \\
0 & 0 & \cdots & 0 & 0 & 0 & 0 & \cdots \\
0 & 0 & \cdots & 0 & 0 & 1-\mu(0) & 0 & \cdots \\
0 & 0 & \cdots & 0 & 0 & 0 & 1-\mu(1) & \cdots \\
\cdots & \cdots & \cdots & \cdots & \cdots & \cdots & \cdots & \cdots \\
0 & 0 & \cdots & 0 & 0 & 0 & 0 & \cdots \\
\lambda(0) & \lambda(1) & \cdots & \lambda(n-1_1) & 1 & 0 & 0 & \cdots \\
0 & 0 & \cdots & 0 & 0 & 0 & 0 & \cdots \\
0 & 0 & \cdots & 0 & 0 & 0 & 0 & \cdots \\
\cdots & \cdots & \cdots & \cdots & \cdots & \cdots & \cdots & \cdots \\
0 & 0 & \cdots & 0 & 0 & 0 & 0 & \cdots
\end{bmatrix}
$$

$$
\begin{bmatrix}
\mu(n_2-1) & 1 & 0 & 0 & \cdots & 0 & 0 \\
0 & 0 & 0 & 0 & \cdots & 0 & 0 \\
0 & 0 & 0 & 0 & \cdots & 0 & 0 \\
\cdots & \cdots & \cdots & \cdots & \cdots & \cdots & \cdots \\
0 & 0 & 0 & 0 & \cdots & 0 & 0 \\
0 & 0 & \rho(0) & \rho(1) & \cdots & \rho(n_3-1) & 1 \\
0 & 0 & 0 & 0 & \cdots & 0 & 0 \\
0 & 0 & 0 & 0 & \cdots & 0 & 0 \\
\cdots & \cdots & \cdots & \cdots & \cdots & \cdots & \cdots \\
1-\mu(n_2-1) & 0 & 0 & 0 & \cdots & 0 & 0 \\
0 & 0 & 0 & 0 & \cdots & 0 & 0 \\
0 & 0 & 1-\rho(0) & 0 & \cdots & 0 & 0 \\
0 & 0 & 0 & 1-\rho(1) & \cdots & 0 & 0 \\
\cdots & \cdots & \cdots & \cdots & \cdots & \cdots & \cdots \\
0 & 0 & 0 & 0 & \cdots & 1-\rho(n_3-1) & 0
\end{bmatrix}.
$$

5.3.2 Stability Proving of Instantaneous Availability

According to features of matrix B and initial value $P(0)$, the following conclusions are set up.

Theorem 5.10: The status of systems (5.8 and 5.9) only exists in finite set C, in which,

$$C = \left\{ \begin{array}{l} Z = \left(z_1, z_2, ..., z_{n_1+n_2+n_3+3}\right)^T \mid \delta_{n_1+n_2+n_3+3,0} Z = 1, 0 \le z_i \le 1, \\ i = 1, 2, ..., n_1 + n_2 + n_3 + 3 \end{array} \right\}.$$

Prove: Because $\delta_{n_1+n_2+n_3+3,0} P(k+1) = \delta_{n_1+n_2+n_3+3,0} BP(k)$, $k = 0, 1, 2, ...,$ for matrix B, obviously,

$$\delta_{n_1+n_2+n_3+3,0} B = \delta_{n_1+n_2+n_3+3,0}.$$

Thus,

$$\delta_{n_1+n_2+n_3+3,0} P(k+1) = \delta_{n_1+n_2+n_3+3,0} P(k) = ... = \delta_{n_1+n_2+n_3+3,0} P(0) = 1.$$

Then, for arbitrary nonnegative integer k, the status $P(k) = \left(p_1, p_2, ..., p_{m+n_2+2}\right)^T$ satisfies

$$\delta_{n_1+n_2+n_3+3,0} P(k) = \sum_{i=1}^{n_1+n_2+n_3+3} p_i = 1. \tag{5.11}$$

According to definitions of failure rate, repair rate, and logistic delay time,

$$0 \le \lambda(k) \le 1 \quad k = 0, 1, 2, ..., n_1$$

$$0 \le \mu(k) \le 1 \quad k = 0, 1, 2, ..., n_2$$

$$0 \le \rho(k) \le 1 \quad k = 0, 1, 2, ..., n_3.$$

Then, arbitrary element in matrix B satisfies:

$$b_{i,j} \ge 0, \quad i, j = 1, 2, ..., n_1 + n_2 + n_3 + 3.$$

Therefore,

$$p_i \ge 0, \quad p_i \le 1, \quad i = 1, 2, ..., n_1 + n_2 + n_3 + 3.$$

The theorem is proven. For systems (5.7 and 5.8), the following formula can be obtained:

$$P(k) = B^k \left(\delta_{1,n_1+n_2+n_3+2}\right)^T \quad k = 0, 1, 2, ... \tag{5.12}$$

In other words, it is the status solution of systems (5.7 and 5.8).

Theorem 5.11: All characteristic roots of matrix B are in a unit circle; that is, $\rho(B) \leq 1$.

Prove: Matrix

$$
B - \gamma I =
\begin{bmatrix}
-\gamma & 0 & \cdots & 0 & 0 & \mu(0) & \mu(1) & \cdots \\
1-\lambda(0) & -\gamma & \cdots & 0 & 0 & 0 & 0 & \cdots \\
0 & 1-\lambda(1) & \cdots & 0 & 0 & 0 & 0 & \cdots \\
\cdots & \cdots & \cdots & \cdots & \cdots & \cdots & \cdots & \cdots \\
0 & 0 & \cdots & 1-\lambda(n_1-1) & -\gamma & 0 & 0 & \cdots \\
0 & 0 & \cdots & 0 & 0 & -\gamma & 0 & \cdots \\
0 & 0 & \cdots & 0 & 0 & 1-\mu(0) & -\gamma & \cdots \\
0 & 0 & \cdots & 0 & 0 & 0 & 1-\mu(1) & \cdots \\
\cdots & \cdots & \cdots & \cdots & \cdots & \cdots & \cdots & \cdots \\
0 & 0 & \cdots & 0 & 0 & 0 & 0 & \cdots \\
\lambda(0) & \lambda(1) & \cdots & \lambda(n_1-1) & 1 & 0 & 0 & \cdots \\
0 & 0 & \cdots & 0 & 0 & 0 & 0 & \cdots \\
0 & 0 & \cdots & 0 & 0 & 0 & 0 & \cdots \\
\cdots & \cdots & \cdots & \cdots & \cdots & \cdots & \cdots & \cdots \\
0 & 0 & \cdots & 0 & 0 & 0 & 0 & \cdots \\
\end{bmatrix}
$$

$$
\begin{bmatrix}
\mu(n_2-1) & 1 & 0 & 0 & \cdots & 0 & 0 \\
0 & 0 & 0 & 0 & \cdots & 0 & 0 \\
0 & 0 & 0 & 0 & \cdots & 0 & 0 \\
\cdots & \cdots & \cdots & \cdots & \cdots & \cdots & \cdots \\
0 & 0 & 0 & 0 & \cdots & 0 & 0 \\
0 & 0 & \rho(0) & \rho(1) & \cdots & \rho(n_3-1) & 1 \\
0 & 0 & 0 & 0 & \cdots & 0 & 0 \\
0 & 0 & 0 & 0 & \cdots & 0 & 0 \\
\cdots & \cdots & \cdots & \cdots & \cdots & \cdots & \cdots \\
1-\mu(n_2-1) & -\gamma & 0 & 0 & \cdots & 0 & 0 \\
0 & 0 & -\gamma & 0 & \cdots & 0 & 0 \\
0 & 0 & 1-\rho(0) & -\gamma & \cdots & 0 & 0 \\
0 & 0 & 0 & 1-\rho(1) & \cdots & 0 & 0 \\
\cdots & \cdots & \cdots & \cdots & \cdots & \cdots & \cdots \\
0 & 0 & 0 & 0 & \cdots & 1-\rho(n_3-1) & -\gamma \\
\end{bmatrix},
$$

when $\forall \gamma \in \{x \mid |x| > 1\}$, $\gamma I - B$ is a strictly diagonally dominant matrix [83]. Therefore, $\gamma I - B$ is nonsingular; that is, $\rho(B) \le 1$.

Theorem 5.12: Rank $(I - B) = n_1 + n_2 + n_3 + 2$.

Prove: Matrix

$$
B - I =
\begin{bmatrix}
-1 & 0 & \cdots & 0 & 0 & \mu(0) & \mu(1) & \cdots \\
1-\lambda(0) & -1 & \cdots & 0 & 0 & 0 & 0 & \cdots \\
0 & 1-\lambda(1) & \cdots & 0 & 0 & 0 & 0 & \cdots \\
\cdots & \cdots & \cdots & \cdots & \cdots & \cdots & \cdots & \cdots \\
0 & 0 & \cdots & 1-\lambda(n_1-1) & -1 & 0 & 0 & \cdots \\
0 & 0 & \cdots & 0 & 0 & -1 & 0 & \cdots \\
0 & 0 & \cdots & 0 & 0 & 1-\mu(0) & -1 & \cdots \\
0 & 0 & \cdots & 0 & 0 & 0 & 1-\mu(1) & \cdots \\
\cdots & \cdots & \cdots & \cdots & \cdots & \cdots & \cdots & \cdots \\
0 & 0 & \cdots & 0 & 0 & 0 & 0 & \cdots \\
\lambda(0) & \lambda(1) & \cdots & \lambda(n_1-1) & 1 & 0 & 0 & \cdots \\
0 & 0 & \cdots & 0 & 0 & 0 & 0 & \cdots \\
0 & 0 & \cdots & 0 & 0 & 0 & 0 & \cdots \\
\cdots & \cdots & \cdots & \cdots & \cdots & \cdots & \cdots & \cdots \\
0 & 0 & \cdots & 0 & 0 & 0 & 0 & \cdots
\end{bmatrix}
$$

$$
\begin{bmatrix}
\mu(n_2-1) & 1 & 0 & 0 & \cdots & 0 & 0 \\
0 & 0 & 0 & 0 & \cdots & 0 & 0 \\
0 & 0 & 0 & 0 & \cdots & 0 & 0 \\
\cdots & \cdots & \cdots & \cdots & \cdots & \cdots & \cdots \\
0 & 0 & 0 & 0 & \cdots & 0 & 0 \\
0 & 0 & \rho(0) & \rho(1) & \cdots & \rho(n_3-1) & 1 \\
0 & 0 & 0 & 0 & \cdots & 0 & 0 \\
0 & 0 & 0 & 0 & \cdots & 0 & 0 \\
\cdots & \cdots & \cdots & \cdots & \cdots & \cdots & \cdots \\
1-\mu(n_2-1) & -1 & 0 & 0 & \cdots & 0 & 0 \\
0 & 0 & -1 & 0 & \cdots & 0 & 0 \\
0 & 0 & 1-\rho(0) & -1 & \cdots & 0 & 0 \\
0 & 0 & 0 & 1-\rho(1) & \cdots & 0 & 0 \\
\cdots & \cdots & \cdots & \cdots & \cdots & \cdots & \cdots \\
0 & 0 & 0 & 0 & \cdots & 1-\rho(n_3-1) & -1
\end{bmatrix}.
$$

Obviously, the sum of elements in all columns of matrix $I-B$ equals 0. Therefore, rank $(I-B) \le n_1 + n_2 + n_3 + 2$. Besides, for matrix $I-B$, the principal minor of order $n_1 + n_2 + n_3 + 2$ of full rank is $(I-B)_{(1,1)}$.

$$
=
\begin{bmatrix}
-1 & 0 & \cdots & 0 & 0 & 0 & 0 & \cdots \\
1-\lambda(1) & -1 & \cdots & 0 & 0 & 0 & 0 & \cdots \\
0 & 1-\lambda(2) & \cdots & 0 & 0 & 0 & 0 & \cdots \\
\cdots & \cdots & \cdots & \cdots & \cdots & \cdots & \cdots & \cdots \\
0 & 0 & \cdots & 1-\lambda(n_1-1) & -1 & 0 & 0 & \cdots \\
0 & 0 & \cdots & 0 & 0 & -1 & 0 & \cdots \\
0 & 0 & \cdots & 0 & 0 & 1-\mu(0) & -1 & \cdots \\
0 & 0 & \cdots & 0 & 0 & 0 & 1-\mu(1) & \cdots \\
\cdots & \cdots & \cdots & \cdots & \cdots & \cdots & \cdots & \cdots \\
0 & 0 & \cdots & 0 & 0 & 0 & 0 & \cdots \\
\lambda(1) & \lambda(2) & \cdots & \lambda(n_1-1) & 1 & 0 & 0 & \cdots \\
0 & 0 & \cdots & 0 & 0 & 0 & 0 & \cdots \\
0 & 0 & \cdots & 0 & 0 & 0 & 0 & \cdots \\
\cdots & \cdots & \cdots & \cdots & \cdots & \cdots & \cdots & \cdots \\
0 & 0 & \cdots & 0 & 0 & 0 & 0 & \cdots
\end{bmatrix}
$$

$$
\begin{bmatrix}
0 & 1 & 0 & 0 & \cdots & 0 & 0 \\
0 & 0 & 0 & 0 & \cdots & 0 & 0 \\
0 & 0 & 0 & 0 & \cdots & 0 & 0 \\
\cdots & \cdots & \cdots & \cdots & \cdots & \cdots & \cdots \\
0 & 0 & 0 & 0 & \cdots & 0 & 0 \\
0 & 0 & \rho(0) & \rho(1) & \cdots & \rho(n_3-1) & 1 \\
0 & 0 & 0 & 0 & \cdots & 0 & 0 \\
0 & 0 & 0 & 0 & \cdots & 0 & 0 \\
\cdots & \cdots & \cdots & \cdots & \cdots & \cdots & \cdots \\
1-\mu(n_2-1) & -1 & 0 & 0 & \cdots & 0 & 0 \\
0 & 0 & -1 & 0 & \cdots & 0 & 0 \\
0 & 0 & 1-\rho(0) & -1 & \cdots & 0 & 0 \\
0 & 0 & 0 & 1-\rho(1) & \cdots & 0 & 0 \\
\cdots & \cdots & \cdots & \cdots & \cdots & \cdots & \cdots \\
0 & 0 & 0 & 0 & \cdots & 1-\rho(n_3-1) & -1
\end{bmatrix}.
$$

Thus, rank $(I-B) \geq n_1 + n_2 + n_3 + 2$. Therefore, rank $(I-B) = n_1 + n_2 + 1$.

Theorem 5.13: Matrix B has characteristic root $\gamma = 1$, which is a simple root. If $X_0 = (x_1, x_2, ..., x_{m_1+n_2+n_3+3})^T$ stands for the feature vector of $\gamma = 1$ on C, the value of X_0 is unique and satisfies

$$x_1 = x_{m+2} = x_{m_1+n_2+3} = \frac{1}{D}$$

$$x_i = \frac{1}{D} \prod_{j=0}^{i-2} \left[1 - \lambda(j)\right], \quad i = 2,3,...,n_1 + 1$$

$$x_{m+i+1} = \frac{1}{D} \prod_{j=0}^{i-2} \left[1 - \mu(j)\right], \quad i = 2,3,...,n_2 + 1$$

$$x_{m_1+n_2+i+1} = \frac{1}{D} \prod_{j=0}^{i-2} \left[1 - \rho(j)\right], \quad i = 2,3,...,n_3 + 1.$$

In the formula,

$$D = 3 + \sum_{i=2}^{n_1+1} \left\{ \prod_{j=0}^{i-2} \left[1 - \lambda(j)\right] \right\} + \sum_{i=2}^{n_2+1} \left\{ \prod_{j=0}^{i-2} \left[1 - \mu(j)\right] \right\}$$

$$+ \sum_{i=2}^{n_3+1} \left\{ \prod_{j=0}^{i-2} \left[1 - \rho(j)\right] \right\}.$$

Prove: According to Theorem 5.12, $\gamma = 1$ is the simple characteristic root of matrix B, which is equivalent to

$$\frac{d|B - \gamma I|}{d\gamma}\Big|_{\gamma=1} \neq 0. \tag{5.13}$$

Then, Formula 5.13 is proven:

$$\frac{d|B - \gamma I|}{d\gamma}\Big|_{\gamma=1}$$

$$
=\begin{bmatrix}
-1 & 0 & \cdots & 0 & 0 & \mu(0) & \mu(1) & \cdots & \mu(n_2-1) & 1 & 0 & 0 & \cdots & 0 & 0 \\
0 & -1 & \cdots & 0 & 0 & 0 & 0 & \cdots & 0 & 0 & 0 & 0 & \cdots & 0 & 0 \\
0 & 1-\lambda(1) & \cdots & 0 & 0 & 0 & 0 & \cdots & 0 & 0 & 0 & 0 & \cdots & 0 & 0 \\
\cdots & \cdots & \cdots & \cdots & \cdots & \cdots & \cdots & \cdots & \cdots & \cdots & \cdots & \cdots & \cdots & \cdots & \cdots \\
0 & 0 & \cdots & 1-\lambda(n_1-1) & -1 & 0 & 0 & \cdots & 0 & 0 & 0 & 0 & \cdots & 0 & 0 \\
0 & 0 & \cdots & 0 & 0 & -1 & 0 & \cdots & 0 & 0 & \rho(0) & \rho(1) & \cdots & \rho(n_3-1) & 1 \\
0 & 0 & \cdots & 0 & 0 & 1-\mu(0) & -1 & \cdots & 0 & 0 & 0 & 0 & \cdots & 0 & 0 \\
0 & 0 & \cdots & 0 & 0 & 0 & 1-\mu(1) & \cdots & 0 & 0 & 0 & 0 & \cdots & 0 & 0 \\
\cdots & \cdots & \cdots & \cdots & \cdots & \cdots & \cdots & \cdots & \cdots & \cdots & \cdots & \cdots & \cdots & \cdots & \cdots \\
0 & 0 & \cdots & 0 & 0 & 0 & 0 & \cdots & 1-\mu(n_2-1) & -1 & 0 & 0 & \cdots & 0 & 0 \\
0 & \lambda(1) & \cdots & \lambda(n_1-1) & 1 & 0 & 0 & \cdots & 0 & 0 & -1 & 0 & \cdots & 0 & 0 \\
0 & 0 & \cdots & 0 & 0 & 0 & 0 & \cdots & 0 & 0 & 1-\rho(0) & -1 & \cdots & 0 & 0 \\
0 & 0 & \cdots & 0 & 0 & 0 & 0 & \cdots & 0 & 0 & 0 & 1-\rho(1) & \cdots & 0 & 0 \\
\cdots & \cdots & \cdots & \cdots & \cdots & \cdots & \cdots & \cdots & \cdots & \cdots & \cdots & \cdots & \cdots & \cdots & \cdots \\
0 & 0 & \cdots & 0 & 0 & 0 & 0 & \cdots & 0 & 0 & 0 & 0 & \cdots & 1-\rho(n_3-1) & -1
\end{bmatrix}
$$

$$
+\begin{bmatrix}
-1 & 0 & \cdots & 0 & 0 & \mu(0) & \mu(1) & \cdots \\
1-\lambda(0) & -1 & \cdots & 0 & 0 & 0 & 0 & \cdots \\
0 & 0 & \cdots & 0 & 0 & 0 & 0 & \cdots \\
\cdots & \cdots & \cdots & \cdots & \cdots & \cdots & \cdots & \cdots \\
0 & 0 & \cdots & 1-\lambda(n_1-1) & -1 & 0 & 0 & \cdots \\
0 & 0 & \cdots & 0 & 0 & -1 & 0 & \cdots \\
0 & 0 & \cdots & 0 & 0 & 1-\mu(0) & -1 & \cdots \\
0 & 0 & \cdots & 0 & 0 & 0 & 1-\mu(1) & \cdots \\
\cdots & \cdots & \cdots & \cdots & \cdots & \cdots & \cdots & \cdots \\
0 & 0 & \cdots & 0 & 0 & 0 & 0 & \cdots \\
\lambda(0) & 0 & \cdots & \lambda(n_1-1) & 1 & 0 & 0 & \cdots \\
0 & 0 & \cdots & 0 & 0 & 0 & 0 & \cdots \\
0 & 0 & \cdots & 0 & 0 & 0 & 0 & \cdots \\
\cdots & \cdots & \cdots & \cdots & \cdots & \cdots & \cdots & \cdots \\
0 & 0 & \cdots & 0 & 0 & 0 & 0 & \cdots
\end{bmatrix}
$$

$$
\begin{bmatrix}
\mu(n_2-1) & 1 & 0 & 0 & \cdots & 0 & 0 \\
0 & 0 & 0 & 0 & \cdots & 0 & 0 \\
0 & 0 & 0 & 0 & \cdots & 0 & 0 \\
\cdots & \cdots & \cdots & \cdots & \cdots & \cdots & \cdots \\
0 & 0 & 0 & 0 & \cdots & 0 & 0 \\
0 & 0 & \rho(0) & \rho(1) & \cdots & \rho(n_3-1) & 1 \\
0 & 0 & 0 & 0 & \cdots & 0 & 0 \\
0 & 0 & 0 & 0 & \cdots & 0 & 0 \\
\cdots & \cdots & \cdots & \cdots & \cdots & \cdots & \cdots \\
1-\mu(n_2-1) & -1 & 0 & 0 & \cdots & 0 & 0 \\
0 & 0 & -1 & 0 & \cdots & 0 & 0 \\
0 & 0 & 1-\rho(0) & -1 & \cdots & 0 & 0 \\
0 & 0 & 0 & 1-\rho(1) & \cdots & 0 & 0 \\
\cdots & \cdots & \cdots & \cdots & \cdots & \cdots & \cdots \\
0 & 0 & 0 & 0 & \cdots & 1-\rho(n_3-1) & -1
\end{bmatrix}
+\cdots
$$

$$
+ \begin{bmatrix}
-1 & 0 & \cdots & 0 & 0 & \mu(0) & \mu(1) & \cdots \\
1-\lambda(0) & -1 & \cdots & 0 & 0 & 0 & 0 & \cdots \\
0 & 1-\lambda(1) & \cdots & 0 & 0 & 0 & 0 & \cdots \\
\cdots & \cdots & \cdots & \cdots & \cdots & \cdots & \cdots & \cdots \\
0 & 0 & \cdots & 1-\lambda(n_1-1) & -1 & 0 & 0 & \cdots \\
0 & 0 & \cdots & 0 & 0 & -1 & 0 & \cdots \\
0 & 0 & \cdots & 0 & 0 & 1-\mu(0) & -1 & \cdots \\
0 & 0 & \cdots & 0 & 0 & 0 & 1-\mu(1) & \cdots \\
\cdots & \cdots & \cdots & \cdots & \cdots & \cdots & \cdots & \cdots \\
0 & 0 & \cdots & 0 & 0 & 0 & 0 & \cdots \\
\lambda(0) & \lambda(1) & \cdots & \lambda(n_1-1) & 0 & 0 & 0 & \cdots \\
0 & 0 & \cdots & 0 & 0 & 0 & 0 & \cdots \\
0 & 0 & \cdots & 0 & 0 & 0 & 0 & \cdots \\
\cdots & \cdots & \cdots & \cdots & \cdots & \cdots & \cdots & \cdots \\
0 & 0 & \cdots & 0 & 0 & 0 & 0 & \cdots
\end{bmatrix}
$$

$$
\begin{bmatrix}
\mu(n_2-1) & 1 & 0 & 0 & \cdots & 0 & 0 \\
0 & 0 & 0 & 0 & \cdots & 0 & 0 \\
0 & 0 & 0 & 0 & \cdots & 0 & 0 \\
\cdots & \cdots & \cdots & \cdots & \cdots & \cdots & \cdots \\
0 & 0 & 0 & 0 & \cdots & 0 & 0 \\
0 & 0 & \rho(0) & \rho(1) & \cdots & \rho(n_3-1) & 1 \\
0 & 0 & 0 & 0 & \cdots & 0 & 0 \\
0 & 0 & 0 & 0 & \cdots & 0 & 0 \\
\cdots & \cdots & \cdots & \cdots & \cdots & \cdots & \cdots \\
1-\mu(n_2-1) & -1 & 0 & 0 & \cdots & 0 & 0 \\
0 & 0 & -1 & 0 & \cdots & 0 & 0 \\
0 & 0 & 1-\rho(0) & -1 & \cdots & 0 & 0 \\
0 & 0 & 0 & 1-\rho(1) & \cdots & 0 & 0 \\
\cdots & \cdots & \cdots & \cdots & \cdots & \cdots & \cdots \\
0 & 0 & 0 & 0 & \cdots & 1-\rho(n_3-1) & -1
\end{bmatrix} + \cdots
$$

$$
+\left[\begin{array}{ccccccccc}
-1 & 0 & \cdots & 0 & 0 & \mu(0) & \mu(1) & \cdots \\
1-\lambda(0) & -1 & \cdots & 0 & 0 & 0 & 0 & \cdots \\
0 & 1-\lambda(1) & \cdots & 0 & 0 & 0 & 0 & \cdots \\
\cdots & \cdots & \cdots & \cdots & \cdots & \cdots & \cdots & \cdots \\
0 & 0 & \cdots & 1-\lambda(n_1-1) & -1 & 0 & 0 & \cdots \\
0 & 0 & \cdots & 0 & 0 & -1 & 0 & \cdots \\
0 & 0 & \cdots & 0 & 0 & 1-\mu(0) & -1 & \cdots \\
0 & 0 & \cdots & 0 & 0 & 0 & 1-\mu(1) & \cdots \\
\cdots & \cdots & \cdots & \cdots & \cdots & \cdots & \cdots & \cdots \\
0 & 0 & \cdots & 0 & 0 & 0 & 0 & \cdots \\
\lambda(0) & \lambda(1) & \cdots & \lambda(n_1-1) & 1 & 0 & 0 & \cdots \\
0 & 0 & \cdots & 0 & 0 & 0 & 0 & \cdots \\
0 & 0 & \cdots & 0 & 0 & 0 & 0 & \cdots \\
\cdots & \cdots & \cdots & \cdots & \cdots & \cdots & \cdots & \cdots \\
0 & 0 & \cdots & 0 & 0 & 0 & 0 & \cdots
\end{array}\right.
$$

$$
\left.\begin{array}{ccccccc}
\mu(n_2) & 1 & 0 & 0 & \cdots & 0 & 0 \\
0 & 0 & 0 & 0 & \cdots & 0 & 0 \\
0 & 0 & 0 & 0 & \cdots & 0 & 0 \\
\cdots & \cdots & \cdots & \cdots & \cdots & \cdots & \cdots \\
0 & 0 & 0 & 0 & \cdots & 0 & 0 \\
0 & 0 & \rho(0) & \rho(1) & \cdots & \rho(n_3-1) & 0 \\
0 & 0 & 0 & 0 & \cdots & 0 & 0 \\
0 & 0 & 0 & 0 & \cdots & 0 & 0 \\
\cdots & \cdots & \cdots & \cdots & \cdots & \cdots & \cdots \\
1-\mu(n_2-1) & -1 & 0 & 0 & \cdots & 0 & 0 \\
0 & 0 & -1 & 0 & \cdots & 0 & 0 \\
0 & 0 & 1-\rho(0) & -1 & \cdots & 0 & 0 \\
0 & 0 & 0 & 1-\rho(1) & \cdots & 0 & 0 \\
\cdots & \cdots & \cdots & \cdots & \cdots & \cdots & \cdots \\
0 & 0 & 0 & 0 & \cdots & 1-\rho(n_3-1) & -1
\end{array}\right]
$$

$$
=(-1)^{m_1+m_2+m_3+2}\left\{\begin{array}{l}
-3+[\lambda(0)-1]+\ldots+[\lambda(n_1-1)-1]+[\mu(0)-1]+\ldots \\
+[\mu(n_1-1)-1]+[\rho(0)-1]+\ldots+[\rho(n_1-1)-1]
\end{array}\right\} \neq 0.
$$

Thus, $\gamma = 1$ is the simple characteristic root of matrix B.

Assuming X_0 stands for the feature vector of $\gamma = 1$ on C. Obviously, $(B - I) X_0 = 0$; that is,

$$
\begin{bmatrix}
-1 & 0 & \cdots & 0 & 0 & \mu(0) & \mu(1) & \cdots & \mu(n_2-1) & 1 & 0 & 0 & \cdots & 0 & 0 \\
1-\lambda(0) & -1 & \cdots & 0 & 0 & 0 & 0 & \cdots & 0 & 0 & 0 & 0 & \cdots & 0 & 0 \\
0 & 1-\lambda(1) & \cdots & 0 & 0 & 0 & 0 & \cdots & 0 & 0 & 0 & 0 & \cdots & 0 & 0 \\
\cdots & \cdots & \cdots & \cdots & \cdots & \cdots & \cdots & \cdots & \cdots & \cdots & \cdots & \cdots & \cdots & \cdots & \cdots \\
0 & 0 & \cdots & 1-\lambda(n_1-1) & -1 & 0 & 0 & \cdots & 0 & 0 & 0 & 0 & \cdots & 0 & 0 \\
0 & 0 & \cdots & 0 & 0 & -1 & 0 & \cdots & 0 & 0 & \rho(0) & \rho(1) & \cdots & \rho(n_3-1) & 1 \\
0 & 0 & \cdots & 0 & 0 & 1-\mu(0) & -1 & \cdots & 0 & 0 & 0 & 0 & \cdots & 0 & 0 \\
0 & 0 & \cdots & 0 & 0 & 0 & 1-\mu(1) & \cdots & 0 & 0 & 0 & 0 & \cdots & 0 & 0 \\
\cdots & \cdots & \cdots & \cdots & \cdots & \cdots & \cdots & \cdots & \cdots & \cdots & \cdots & \cdots & \cdots & \cdots & \cdots \\
0 & 0 & \cdots & 0 & 0 & 0 & 0 & \cdots & 1-\mu(n_2-1) & -1 & 0 & 0 & \cdots & 0 & 0 \\
\lambda(0) & \lambda(1) & \cdots & \lambda(n_1-1) & 1 & 0 & 0 & \cdots & 0 & 0 & -1 & 0 & \cdots & 0 & 0 \\
0 & 0 & \cdots & 0 & 0 & 0 & 0 & \cdots & 0 & 0 & 1-\rho(0) & -1 & \cdots & 0 & 0 \\
0 & 0 & \cdots & 0 & 0 & 0 & 0 & \cdots & 0 & 0 & 0 & 1-\rho(1) & \cdots & 0 & 0 \\
\cdots & \cdots & \cdots & \cdots & \cdots & \cdots & \cdots & \cdots & \cdots & \cdots & \cdots & \cdots & \cdots & \cdots & \cdots \\
0 & 0 & \cdots & 0 & 0 & 0 & 0 & \cdots & 0 & 0 & 0 & 0 & \cdots & 1-\rho(n_3-1) & -1
\end{bmatrix} X_0 = 0.
$$

According to Theorem 5.12, the rank of matrix $I - B$ is $n_1 + n_2 + n_3 + 2$. Then,

$$
\begin{cases}
x_2 = \left[1 - \lambda(0)\right]x_1 \\
x_3 = \left[1 - \lambda(0)\right]\left[1 - \lambda(1)\right]x_1 \\
\cdots \\
x_{m_1+1} = \displaystyle\prod_{i=0}^{n_1-1}\left[1 - \lambda(i)\right]x_1 \\
x_{m_1+3} = \left[1 - \mu(0)\right]x_{m_1+2} \\
x_{m_1+4} = \left[1 - \mu(0)\right]\left[1 - \mu(1)\right]x_{m_1+2} \\
\cdots \\
x_{m_1+n_2+2} = \displaystyle\prod_{i=0}^{n_1-1}\left[1 - \mu(i)\right]x_{m_1+2} \\
x_{m_1+n_2+4} = [1 - \rho(0)]x_{m_1+n_2+3} \\
x_{m_1+n_2+5} = [1 - \rho(0)][1 - \rho(1)]x_{m_1+n_2+3} \\
\cdots \\
x_{m_1+n_2+n_3+3} = \displaystyle\prod_{i=0}^{n_3-1}\left[1 - \rho(i)\right]x_{m_1+n_2+3} \\
\displaystyle\sum_{i=1}^{n_2+1}\left[x_{m_1+i+1}\mu(i-1)\right] - x_1 = 0 \\
\displaystyle\sum_{i=1}^{n_3+1}\left[x_{m_1+n_2+2+i}\rho(i-1)\right] - x_{m_1+2} = 0.
\end{cases}
$$

In other words,

$$
x_i = \prod_{j=0}^{i-2}[1 - \lambda(j)]x_1, \quad i = 2, 3, \ldots, n_2 + 1
$$

$$
x_{m_1+i+1} = \prod_{j=0}^{i-2}[1 - \mu(j)]x_{m_1+2}, \quad i = 2, 3, \ldots, n_2 + 1
$$

$$x_{m_1+n_2+i+2} = \prod_{j=0}^{i-2}[1-\rho(j)]x_{m_1+2}, \quad i = 2, 3, \ldots, n_3 + 1$$

$$\sum_{i=1}^{n_2+1}[x_{m_1+i+1}\mu(i-1)] - x_1 = 0$$

$$\sum_{i=1}^{n_3+1}[x_{m_1+n_2+2+i}\rho(i-1)] - x_{m_1+2} = 0.$$

Suppose

$$D_1 = \lambda(0) + \sum_{i=2}^{m_1+1}\left\{\prod_{j=0}^{i-2}[1-\lambda(j)]\lambda(i-1)\right\}$$

$$D_2 = \mu(0) + \sum_{i=2}^{n_2+1}\left\{\prod_{j=0}^{i-2}[1-\mu(j)]\mu(i-1)\right\}$$

$$D_3 = \rho(0) + \sum_{i=2}^{n_3+1}\left\{\prod_{j=0}^{i-2}[1-\rho(j)]\rho(i-1)\right\}.$$

Then,

$$x_1 = D_2 x_{m_1+2}$$

$$x_{m_1+2} = D_2 x_{m_1+n_2+3}.$$

The following conclusion is easily proven:

$$D_1 = D_2 = D_3 = 1.$$

Then,

$$x_1 = x_{m_1+2} = x_{m_1+n_2+3}.$$

Because $X_0 \in C$; that is,

$$\left(1 + \sum_{i=2}^{n_1+1}\left\{\prod_{j=0}^{i-2}[1-\lambda(j)]\right\}\right)x_1 + \left(1 + \sum_{i=2}^{n_2+1}\left\{\prod_{j=0}^{i-2}[1-\mu(j)]\right\}\right)x_1$$

$$+ \left(1 + \sum_{i=2}^{n_3+1}\left\{\prod_{j=0}^{i-2}[1-\rho(j)]\right\}\right)x_1 = 1.$$

Suppose

$$D = 3 + \sum_{i=2}^{n_1+1}\left\{\prod_{j=0}^{i-2}[1-\lambda(j)]\right\}$$

$$+ \sum_{i=2}^{n_2+1}\left\{\prod_{j=0}^{i-2}[1-\mu(j)]\right\}$$

$$+ \sum_{i=2}^{n_3+1}\left\{\prod_{j=0}^{i-2}[1-\rho(j)]\right\}.$$

Then,

$$x_1 = x_{n_1+2} = x_{n_1+n_2+3} = \frac{1}{D}$$

$$x_i = \frac{1}{D}\prod_{j=0}^{i-2}[1-\lambda(j)], \quad i = 2, 3, \ldots, n_1 + 1$$

$$x_{n_1+i+1} = \frac{1}{D}\prod_{j=0}^{i-2}[1-\mu(j)], \quad i = 2, 3, \ldots, n_2 + 1$$

$$x_{n_1+n_2+i+1} = \frac{1}{D}\prod_{j=0}^{i-2}[1-\rho(j)], \quad i = 2, 3, \ldots, n_3 + 1.$$

Theorem 5.14: The characteristic root of matrix B, which corresponds to $\rho(B) = 1$, can only be $\gamma = 1$.

Prove: Suppose that the characteristic root of matrix B that corresponds to $\rho(B) = 1$ is $\gamma = e^{bw}$, in which $w^2 = -1$ and $0 \le b < 2\pi$. Suppose the feature vector corresponding to characteristic root $\gamma = e^{bw}$ of matrix B is $\zeta = \left(x_1, x_2, \ldots, x_{n_1+n_2+n_3+3}\right)^T \ne 0$. Then, $(B - \gamma I)\zeta = 0$.

In other words,

$$
\left\{
\begin{array}{l}
-\gamma x_1 + \displaystyle\sum_{i=1}^{n_2+1} [x_{m_1+i+1}\mu(i-1)] = 0 \\[2ex]
[1-\lambda(0)]x_1 - \gamma x_2 = 0 \\[1ex]
[1-\lambda(1)]x_2 - \gamma x_3 = 0 \\[1ex]
\cdots \\[1ex]
[1-\lambda(n_1-1)]x_{n_1} - \gamma x_{n_1+1} = 0 \\[2ex]
\displaystyle\sum_{i=1}^{n_2+1} [x_{m_1+n_2+2+i}\rho(i-1)] - \gamma x_{m_1+2} = 0 \\[2ex]
[1-\mu(0)]x_{m_1+2} - \gamma x_{m_1+3} = 0 \\[1ex]
[1-\mu(1)]x_{m_1+3} - \gamma x_{m_1+4} = 0 \\[1ex]
\cdots \\[1ex]
[1-\mu(n_2-1)]x_{m_1+n_2+1} - \gamma x_{m_1+n_2+2} = 0 \\[2ex]
\displaystyle\sum_{i=1}^{n_2+1} [x_i\lambda(i-1)] - \gamma x_{m_1+n_2+3} = 0 \\[2ex]
[1-\rho(0)]x_{m_1+n_2+3} - \gamma x_{m_1+n_2+4} = 0 \\[1ex]
[1-\rho(1)]x_{m_1+n_2+4} - \gamma x_{m_1+n_2+5} = 0 \\[1ex]
\cdots \\[1ex]
[1-\rho(n_2-1)]x_{m_1+n_2+2} - \gamma x_{m_1+n_2+3} = 0.
\end{array}
\right.
$$

Then,

$$
x_i = \prod_{j=0}^{i-2}\left[\frac{1-\lambda(j)}{\gamma}\right]x_1, \quad i = 2, 3, \ldots, n_1+1
$$

$$
x_{m_1+i+1} = \prod_{j=0}^{i-2}\left[\frac{1-\mu(j)}{\gamma}\right]x_{m_1+2}, \quad i = 2, 3, \ldots, n_2+1
$$

$$x_{m_1+m_2+i+1} = \prod_{j=0}^{i-2}\left[\frac{1-\rho(j)}{\gamma}\right]x_{m_1+m_2+3}, \quad i = 2, 3, \ldots, n_3+1$$

$$-x_1 + \sum_{i=1}^{n_2+1}\left[x_{m_1+i+1}\frac{\mu(i-1)}{\gamma}\right] = 0$$

$$\sum_{i=1}^{n_2+1}\left[x_{m_1+m_2+2+i}\frac{\rho(i-1)}{\gamma}\right] - x_{m_1+2} = 0$$

$$\sum_{i=1}^{m_1+1}\left[x_i\frac{\lambda(i-1)}{\gamma}\right] - x_{m_1+m_2+3} = 0.$$

Suppose

$$D_1 = \lambda(0)e^{-bw} + \sum_{i=2}^{n_2+1}\left\{\prod_{j=0}^{i-2}[1-\lambda(j)]\lambda(i-1)e^{-biw}\right\}$$

$$D_2 = \mu(0)e^{-bw} + \sum_{i=2}^{n_2+1}\left\{\prod_{j=0}^{i-2}[1-\mu(j)]\mu(i-1)e^{-biw}\right\}$$

$$D_3 = \rho(0)e^{-bw} + \sum_{i=2}^{n_2+1}\left\{\prod_{j=0}^{i-2}[1-\rho(j)]\rho(i-1)e^{-biw}\right\}.$$

Then,

$$x_1 = D_2 x_{m_1+2}$$
$$x_{m_1+2} = D_3 x_{m_1+m_2+3}$$
$$x_{m_1+m_2+3} = D_1 x_1.$$

Thus,

$$(D_2 D_3 D_1 - 1)x_1 = 0.$$

Obviously, $x_1 \neq 0$. Otherwise, $x_i = 0$ and $i = 1, 2, \ldots, n_1 + n_2 + n_3 + 3$, which is contradictory to $\zeta = (x_1, x_2, \ldots, x_{2n+2})^{\mathrm{T}} \neq 0$. Then,

$$D_2 D_3 D_1 = 1$$

because

$$|D_2 D_3 D_1| = |D_2||D_3||D_1| = 1.$$

In the formula,

$$| D_1 | = \left| \lambda(0)e^{-bw} + \sum_{i=2}^{n_1+1} \left\{ \prod_{j=0}^{i-2} [1 - \lambda(j)]\lambda(i-1)e^{-biw} \right\} \right|$$

$$\leq \lambda(0) | e^{-bw} | + \sum_{i=2}^{n_1+1} \left\{ \prod_{j=0}^{i-2} [1 - \lambda(j)]\lambda(i-1) | e^{-biw} | \right\}$$

$$= \lambda(0) + \sum_{i=2}^{n_1+1} \left\{ \prod_{j=0}^{i-2} [1 - \lambda(j)]\lambda(i-1) \right\} = 1$$

$$| D_2 | = \left| \mu(0)e^{-bw} + \sum_{i=2}^{n_2+1} \left\{ \prod_{j=0}^{i-2} [1 - \mu(j)]\mu(i-1)e^{-biw} \right\} \right|$$

$$\leq \mu(0) | e^{-bw} | + \sum_{i=2}^{n_2+1} \left\{ \prod_{j=0}^{i-2} [1 - \mu(j)]\mu(i-1) | e^{-biw} | \right\}$$

$$= \mu(0) + \sum_{i=2}^{n_2+1} \left\{ \prod_{j=0}^{i-2} [1 - \mu(j)]\mu(i-1) \right\} = 1$$

$$| D_3 | = \left| \rho(0)e^{-bw} + \sum_{i=2}^{n_2+1} \left\{ \prod_{j=0}^{i-2} [1 - \rho(j)]\rho(i-1)e^{-biw} \right\} \right|$$

$$\leq \rho(0) | e^{-bw} | + \sum_{i=2}^{n_2+1} \left\{ \prod_{j=0}^{i-2} [1 - \rho(j)]\rho(i-1) | e^{-biw} | \right\}$$

$$= \rho(0) + \sum_{i=2}^{n_2+1} \left\{ \prod_{j=0}^{i-2} [1 - \rho(j)]\rho(i-1) \right\} = 1.$$

According to Reference [84], the necessary and sufficient conditions for establishment of the equal sign in this formula are as follows:

1. All vectors expressed by $\lambda(0)e^{-bw}$ and $\prod_{j=0}^{i-2} [1 - \lambda(j)]\lambda(i-1)e^{-biw}$ $(i = 2, 3, \ldots, n_1 + 1)$ are collinear with the same direction.

2. All vectors expressed by $\mu(0)e^{-bw}$ and $\prod_{j=0}^{i-2} [1 - \mu(j)]\mu(i-1)e^{-biw}$ $(i = 2, 3, \ldots, n_2 + 1)$ are collinear with the same direction.

3. All vectors expressed by $\rho(0)e^{-bw}$ and $\prod_{j=0}^{i-2}[1-\rho(j)]\rho(i-1)e^{-biw}$ $(i = 2,3,\ldots,n_3+1)$

are collinear with the same direction.

According to the value range of $0 \le b < 2\pi$, $b = 0$ can be obtained. Thus, the characteristic root of matrix B corresponding to $\rho(B) = 1$ can only be $\gamma = 1$.

The following conclusions can be drawn in the combination of Theorems 5.11–5.14.

Theorem 5.15: Status solution (5.12) of systems (5.8 through 5.10) is stable and the unique equilibrium point is unique and equal to X_0; that is, the feature vector of characteristic root $\gamma = 1$ of matrix B on C.

Prove: According to Theorem 5.8.6 in Reference [85], the status solution (5.12) of systems (5.8 through 5.10) is stable. Suppose the equilibrium point is X_1, $X_1 \in C$ and $X_1 = BX_1$. Then, X_1 is feature vector of characteristic root $\gamma = 1$ of matrix B on C. Then, $X_1 = X_0$ can be obtained as per Theorem 5.13.

Theorem 5.16: System instantaneous availability (5.10) is stable, and the equilibrium point is unique; that is, the steady-state availability of the system is unique with a value as follows:

$$A = \sum_{i=1}^{n_1+1} x_i = \frac{1}{D}\left(1 + \sum_{i=2}^{n_1+1}\left\{\prod_{j=0}^{i-2}[1-\lambda(j)]\right\}\right). \tag{5.14}$$

In the formula,

$$D = 3 + \sum_{i=2}^{n_1+1}\left\{\prod_{j=0}^{i-2}[1-\lambda(j)]\right\} + \sum_{i=2}^{n_2+1}\left\{\prod_{j=0}^{i-2}[1-\mu(j)]\right\} + \sum_{i=2}^{n_3+1}\left\{\prod_{j=0}^{i-2}[1-\rho(j)]\right\}.$$

Prove: According to Theorem 5.15, the following formula can be obtained:

$$X_0 = \lim_{k\to\infty} P(k).$$

According to Theorem 5.13, the limit

$$A = \lim_{k\to\infty} A(k) = \lim_{k\to\infty}\sigma_{n_1+1,n_2+n_3+2}P(k)$$

$$= \sigma_{n_1+1,n_2+n_3+2}X_0 = \sum_{i=1}^{n_1+1} x_i = \frac{1}{D}\left(1 + \sum_{i=2}^{n_1+1}\left\{\prod_{j=0}^{i-2}[1-\lambda(j)]\right\}\right)$$

exists; that is, the steady-state availability of the system.

Obviously,

$$A = \frac{MTBF}{MTBF + MTTR + MLDT}.$$

In other words, the steady-state availability and inherent availability of system are numerically equal. They will not be differentiated in this chapter.

5.4 Repairable System with Preventive Maintenance Considered

5.4.1 System Instantaneous Availability Model

In this section, a model with finite time restrained is researched. Suppose corrective maintenance time and preventive maintenance time are both limited; that is, there is n_2, $n_3 \in Z^+$, s.t.,

$$\mu_1(n_2) = 1, \mu_2(n_3) = 1.$$

Theorem 5.17: When arbitrary $k_2 > n_2$ and $k_3 > n_3$,

$$P_1(k, k_2) = 0 \text{ and } P_2(k, k_3) = 0.$$

Prove: When arbitrary $k_1 > n_2$,

$$P_1(k, k_2) = P_1(k-1, k_2-1)[1 - \mu_1(k_2 - 1)] = \dots$$

$$= P_1(k - k_2 + n_2, n_2) \prod_{j=n_2}^{k_1-1} [1 - \mu_1(j)] = 0.$$

Similarly, it can be proven that when arbitrary

$$k_3 > n_3, P_2(k, k_3) = 0.$$

For Section 4.2, for convenience, make $n_1 = T_0$ and $P_0(k, k_1) = 0$, in which, $k_1 > n_1$. Therefore, $P_0(k, k_1)$ may not be considered. According to Theorem 5.17, it is also feasible to not research $P_1(k, k_2)$ and $P_2(k, k_3)$, in which, $k_2 > n_2$ and $k_3 > n_3$. Therefore, suppose

$$P(k) = (P_0(k), P_1(k), P_2(k))^T.$$

In the formula,

$$P_0(k) = (P_0(k,0), P_0(k,1), \dots, P_0(k,n_1))^T \in [0,1]^{n_1+1}$$

$$P_1(k) = (P_1(k,0), P_1(k,1), \dots, P_1(k,n_2))^T \in [0,1]^{n_2+1}$$

$$P_2(k) = (P_2(k,0), P_2(k,1), \dots, P_2(k,n_3))^T \in [0,1]^{n_3+1}.$$

Then,

$$P(k + 1) = BP(k), \quad k = 0, 1, 2, \dots \tag{5.15}$$

$$P(0) = (\delta_{1, n_1+n_2+n_3+2})^T. \tag{5.16}$$

System instantaneous availability is as follows:

$$A(k) = \delta_{m_1+1,\, n_2+n_3+2} P(k).$$ (5.17)

In the formula,

$$
B = \begin{bmatrix}
0 & 0 & \cdots & 0 & 0 & \mu_1(0) & \mu_1(0) & \cdots & \mu_1(n_2-1) & 1 & \mu_2(0) & \mu_2(1) & \cdots & \mu_2(n_3-1) & 1 \\
1-\lambda(0) & 0 & \cdots & 0 & 0 & 0 & 0 & \cdots & 0 & 0 & 0 & 0 & \cdots & 0 & 0 \\
0 & 1-\lambda(1) & \cdots & 0 & 0 & 0 & 0 & \cdots & 0 & 0 & 0 & 0 & \cdots & 0 & 0 \\
\cdots & \cdots & \cdots & \cdots & \cdots & \cdots & \cdots & \cdots & \cdots & \cdots & \cdots & \cdots & \cdots & \cdots & \cdots \\
0 & 0 & \cdots & 1-\lambda(n_1-1) & 0 & 0 & 0 & \cdots & 0 & 0 & 0 & 0 & \cdots & 0 & 0 \\
\lambda(0) & \lambda(1) & \cdots & \lambda(n_1-1) & \lambda(n_1) & 0 & 0 & \cdots & 0 & 0 & 0 & 0 & \cdots & 0 & 0 \\
0 & 0 & \cdots & 0 & 0 & 1-\mu_1(0) & 0 & \cdots & 0 & 0 & 0 & 0 & \cdots & 0 & 0 \\
0 & 0 & \cdots & 0 & 0 & 0 & 1-\mu_1(1) & \cdots & 0 & 0 & 0 & 0 & \cdots & 0 & 0 \\
\cdots & \cdots & \cdots & \cdots & \cdots & \cdots & \cdots & \cdots & \cdots & \cdots & \cdots & \cdots & \cdots & \cdots & \cdots \\
0 & 0 & \cdots & 0 & 0 & 0 & 0 & \cdots & 1-\mu_1(n_2-1) & 0 & 0 & 0 & \cdots & 0 & 0 \\
0 & 0 & \cdots & 0 & 1-\lambda(n_1) & 0 & 0 & \cdots & 0 & 0 & 0 & 0 & \cdots & 0 & 0 \\
0 & 0 & \cdots & 0 & 0 & 0 & 0 & \cdots & 0 & 0 & 1-\mu_2(0) & 0 & \cdots & 0 & 0 \\
0 & 0 & \cdots & 0 & 0 & 0 & 0 & \cdots & 0 & 0 & 0 & 1-\mu_2(1) & \cdots & 0 & 0 \\
\cdots & \cdots & \cdots & \cdots & \cdots & \cdots & \cdots & \cdots & \cdots & \cdots & \cdots & \cdots & \cdots & \cdots & \cdots \\
0 & 0 & \cdots & 0 & 0 & 0 & 0 & \cdots & 0 & 0 & 0 & 0 & \cdots & 1-\mu_2(n_3-1) & 0
\end{bmatrix}.
$$

5.4.2 Stability Proving of Instantaneous Availability

According to features of matrix B and initial value $P(0)$, the following conclusions are set up.

Theorem 5.18: The status of systems (5.15 and 5.16) only exists in finite set C, in which,

$$C = \{Z = (z_1, z_2, \ldots, z_{n_1+n_2+n_3+3})^T \mid \delta_{n_1+n_2+n_3+3,0} Z = 1, 0 \le z_i \le 1,$$

$$i = 1, 2, \ldots, n_1 + n_2 + n_3 + 3\}.$$

Prove: Since

$$\delta_{n_1+n_2+n_3+3,0} P(k+1) = \delta_{n_1+n_2+n_3+3,0} BP(k), \quad k = 0, 1, 2, \ldots,$$

For matrix B, obviously,

$$\delta_{n_1+n_2+n_3+3,0} B = \delta_{n_1+n_2+n_3+3,0}.$$

Thus,

$$\delta_{n_1+n_2+n_3+3,0} P(k+1) = \delta_{n_1+n_2+n_3+3,0} P(k) = \ldots = \delta_{n_1+n_2+n_3+3,0} P(0) = 1.$$

Then, for arbitrary nonnegative integer k, the status $P(k) = (p_1, p_2, \ldots, p_{n_1+n_2+2})^T$ satisfies

$$\delta_{n_1+n_2+n_3+3,0} P(k) = \sum_{i=1}^{n_1+n_2+n_3+3} p_i = 1. \tag{5.18}$$

According to definitions of failure rate and repair rate,

$$0 \le \lambda(k) \le 1 \quad k = 0, 1, 2, \ldots, n_1$$

$$0 \le \mu_1(k) \le 1 \quad k = 0, 1, 2, \ldots, n_2$$

$$0 \le \mu_2(k) \le 1 \quad k = 0, 1, 2, \ldots, n_3.$$

Then, arbitrary element in matrix B satisfies:

$$b_{i,j} \ge 0, \quad i, j = 1, 2, \ldots, n_1 + n_2 + n_3 + 3.$$

Then,

$$p_i \ge 0, p_i \le 1, \quad i = 1, 2, \ldots, n_1 + n_2 + n_3 + 3.$$

The theorem is proven. For systems (5.15 and 5.16), the following formula can be obtained:

$$P(k) = B^k (\delta_{1,n_1+n_2+n_3+2})^T \quad k = 0, 1, 2, \ldots \tag{5.19}$$

In other words, it is a status solution of systems (5.15 and 5.16).

Theorem 5.19: Characteristic roots of matrix B are in a unit circle; that is, $\rho(B) \le 1$.

Prove: Matrix

$$
B - \gamma I =
\begin{bmatrix}
-\gamma & 0 & \cdots & 0 & 0 & \mu_1(0) & \mu_1(0) & \cdots \\
1-\lambda(0) & -\gamma & \cdots & 0 & 0 & 0 & 0 & \cdots \\
0 & 1-\lambda(1) & \cdots & 0 & 0 & 0 & 0 & \cdots \\
\cdots & \cdots & \cdots & \cdots & \cdots & \cdots & \cdots & \cdots \\
0 & 0 & \cdots & 1-\lambda(n_1-1) & -\gamma & 0 & 0 & \cdots \\
\lambda(0) & \lambda(1) & \cdots & \lambda(n_1-1) & \lambda(n_1) & -\gamma & 0 & \cdots \\
0 & 0 & \cdots & 0 & 0 & 1-\mu_1(0) & -\gamma & \cdots \\
0 & 0 & \cdots & 0 & 0 & 0 & 1-\mu_1(1) & \cdots \\
\cdots & \cdots & \cdots & \cdots & \cdots & \cdots & \cdots & \cdots \\
0 & 0 & \cdots & 0 & 1-\lambda(n_1) & 0 & 0 & \cdots \\
0 & 0 & \cdots & 0 & 0 & 0 & 0 & \cdots \\
0 & 0 & \cdots & 0 & 0 & 0 & 0 & \cdots \\
\cdots & \cdots & \cdots & \cdots & \cdots & \cdots & \cdots & \cdots \\
0 & 0 & \cdots & 0 & 0 & 0 & 0 & \cdots
\end{bmatrix}
$$

$$
\begin{bmatrix}
\mu_1(n_2-1) & 1 & \mu_2(0) & \mu_2(1) & \cdots & \mu_2(n_3-1) & 1 \\
0 & 0 & 0 & 0 & \cdots & 0 & 0 \\
0 & 0 & 0 & 0 & \cdots & 0 & 0 \\
\cdots & \cdots & \cdots & \cdots & \cdots & \cdots & \cdots \\
0 & 0 & 0 & 0 & \cdots & 0 & 0 \\
0 & 0 & 0 & 0 & \cdots & 0 & 0 \\
0 & 0 & 0 & 0 & \cdots & 0 & 0 \\
0 & 0 & 0 & 0 & \cdots & 0 & 0 \\
\cdots & \cdots & \cdots & \cdots & \cdots & \cdots & \cdots \\
1-\mu_1(n_2-1) & -\gamma & 0 & 0 & \cdots & 0 & 0 \\
0 & 0 & -\gamma & 0 & \cdots & 0 & 0 \\
0 & 0 & 1-\mu_2(0) & -\gamma & \cdots & 0 & 0 \\
0 & 0 & 0 & 1-\mu_2(1) & \cdots & 0 & 0 \\
\cdots & \cdots & \cdots & \cdots & \cdots & \cdots & \cdots \\
0 & 0 & 0 & 0 & \cdots & 1-\mu_2(n_3-1) & -\gamma
\end{bmatrix}.
$$

Obviously, when $\forall \gamma \in \{x| |x|>1\}$, $\gamma I - B$ is strictly a diagonally dominant matrix [83]. Therefore, $\gamma I - B$ is nonsingular; that is, $|\gamma I - B| \neq 0$. So, $\rho(B) \leq 1$.

Theorem 5.20: Rank $(I - B) = n_1 + n_2 + n_3 + 2$.

Prove: Matrix

$$B - I = \begin{bmatrix}
-1 & 0 & \cdots & 0 & 0 & \mu_1(0) & \mu_1(0) & \cdots \\
1-\lambda(0) & -1 & \cdots & 0 & 0 & 0 & 0 & \cdots \\
0 & 1-\lambda(1) & \cdots & 0 & 0 & 0 & 0 & \cdots \\
\cdots & \cdots & \cdots & \cdots & \cdots & \cdots & \cdots & \cdots \\
0 & 0 & \cdots & 1-\lambda(n_1-1) & -1 & 0 & 0 & \cdots \\
\lambda(0) & \lambda(1) & \cdots & \lambda(n_1-1) & \lambda(n_1) & -1 & 0 & \cdots \\
0 & 0 & \cdots & 0 & 0 & 1-\mu_1(0) & -1 & \cdots \\
0 & 0 & \cdots & 0 & 0 & 0 & 1-\mu_1(1) & \cdots \\
\cdots & \cdots & \cdots & \cdots & \cdots & \cdots & \cdots & \cdots \\
0 & 0 & \cdots & 0 & 0 & 0 & 0 & \cdots \\
0 & 0 & \cdots & 0 & 1-\lambda(n_1) & 0 & 0 & \cdots \\
0 & 0 & \cdots & 0 & 0 & 0 & 0 & \cdots \\
0 & 0 & \cdots & 0 & 0 & 0 & 0 & \cdots \\
\cdots & \cdots & \cdots & \cdots & \cdots & \cdots & \cdots & \cdots \\
0 & 0 & \cdots & 0 & 0 & 0 & 0 & \cdots
\end{bmatrix}$$

$$\begin{bmatrix}
\mu_1(n_2-1) & 1 & \mu_2(0) & \mu_2(1) & \cdots & \mu_2(n_3-1) & 1 \\
0 & 0 & 0 & 0 & \cdots & 0 & 0 \\
0 & 0 & 0 & 0 & \cdots & 0 & 0 \\
\cdots & \cdots & \cdots & \cdots & \cdots & \cdots & \cdots \\
0 & 0 & 0 & 0 & \cdots & 0 & 0 \\
0 & 0 & 0 & 0 & \cdots & 0 & 0 \\
0 & 0 & 0 & 0 & \cdots & 0 & 0 \\
0 & 0 & 0 & 0 & \cdots & 0 & 0 \\
\cdots & \cdots & \cdots & \cdots & \cdots & \cdots & \cdots \\
1-\mu_1(n_2-1) & -1 & 0 & 0 & \cdots & 0 & 0 \\
0 & 0 & -1 & 0 & \cdots & 0 & 0 \\
0 & 0 & 1-\mu_2(0) & -1 & \cdots & 0 & 0 \\
0 & 0 & 0 & 1-\mu_2(1) & \cdots & 0 & 0 \\
\cdots & \cdots & \cdots & \cdots & \cdots & \cdots & \cdots \\
0 & 0 & 0 & 0 & \cdots & 1-\mu_2(n_3-1) & -1
\end{bmatrix}.$$

$$
=\begin{bmatrix}
-1 & 0 & \cdots & 0 & 0 & 0 & 0 & \cdots \\
1-\lambda(1) & -1 & \cdots & 0 & 0 & 0 & 0 & \cdots \\
0 & 1-\lambda(2) & \cdots & 0 & 0 & 0 & 0 & \cdots \\
\cdots & \cdots & \cdots & \cdots & \cdots & \cdots & \cdots & \cdots \\
0 & 0 & \cdots & 1-\lambda(n_1-1) & -1 & 0 & 0 & \cdots \\
\lambda(1) & \lambda(2) & \cdots & \lambda(n_1-1) & \lambda(n_1) & -1 & 0 & \cdots \\
0 & 0 & \cdots & 0 & 0 & 1-\mu_1(0) & -1 & \cdots \\
0 & 0 & \cdots & 0 & 0 & 0 & 1-\mu_1(1) & \cdots \\
\cdots & \cdots & \cdots & \cdots & \cdots & \cdots & \cdots & \cdots \\
0 & 0 & \cdots & 0 & 0 & 0 & 0 & \cdots \\
0 & 0 & \cdots & 0 & 1-\lambda(n_1) & 0 & 0 & \cdots \\
0 & 0 & \cdots & 0 & 0 & 0 & 0 & \cdots \\
0 & 0 & \cdots & 0 & 0 & 0 & 0 & \cdots \\
\cdots & \cdots & \cdots & \cdots & \cdots & \cdots & \cdots & \cdots \\
0 & 0 & \cdots & 0 & 0 & 0 & 0 & \cdots
\end{bmatrix}
$$

$$
\begin{bmatrix}
0 & 0 & 0 & 0 & \cdots & 0 & 0 \\
0 & 0 & 0 & 0 & \cdots & 0 & 0 \\
0 & 0 & 0 & 0 & \cdots & 0 & 0 \\
\cdots & \cdots & \cdots & \cdots & \cdots & \cdots & \cdots \\
0 & 0 & 0 & 0 & \cdots & 0 & 0 \\
0 & 0 & 0 & 0 & \cdots & 0 & 0 \\
0 & 0 & 0 & 0 & \cdots & 0 & 0 \\
0 & 0 & 0 & 0 & \cdots & 0 & 0 \\
\cdots & \cdots & \cdots & \cdots & \cdots & \cdots & \cdots \\
1-\mu_1(n_2-1) & -1 & 0 & 0 & \cdots & 0 & 0 \\
0 & 0 & -1 & 0 & \cdots & 0 & 0 \\
0 & 0 & 1-\mu_2(0) & -1 & \cdots & 0 & 0 \\
0 & 0 & 0 & 1-\mu_2(1) & \cdots & 0 & 0 \\
\cdots & \cdots & \cdots & \cdots & \cdots & \cdots & \cdots \\
0 & 0 & 0 & 0 & \cdots & 1-\mu_2(n_3-1) & -1
\end{bmatrix}.
$$

Thus, rank $(B-I) \geq n_1 + n_2 + n_3 + 2$. Therefore, rank $(I-B) = n_1 + n_2 + n_3 + 2$.

Theorem 5.21: Matrix B has a characteristic root $\gamma = 1$, which is a simple root. If $X_0 = (x_1, x_2, \ldots, x_{m+n_2+n_3+3})^T$ stands for the feature vector of $\gamma = 1$ on C, the value of X_0 is unique and satisfies

$$
x_1 = \frac{1}{D}
$$

$$x_i = \frac{1}{D} \prod_{j=0}^{i-2} [1 - \lambda(j)], \quad i = 2, 3, \ldots, n_1 + 1$$

$$x_{m_1+2} = \frac{D_1}{D}$$

$$x_{m_1+i+1} = \frac{1}{D} \prod_{j=0}^{i-2} [1 - \mu_1(j)], \quad i = 2, 3, \ldots, n_2 + 1$$

$$x_{m_1+n_2+3} = \frac{D_2}{D}$$

$$x_{m_1+n_2+i+1} = \frac{1}{D} \prod_{j=0}^{i-2} [1 - \mu_2(j)], \quad i = 2, 3, \ldots, n_3 + 1.$$

In the formula,

$$D_1 = \lambda(0) + \sum_{i=2}^{n_1+1} \left\{ \prod_{j=0}^{i-2} [1 - \lambda(j)]\lambda(i-1) \right\}$$

$$D_2 = \prod_{j=0}^{m_1-1} [1 - \lambda(j)][1 - \lambda(n_1)]$$

$$D = 2 + \sum_{i=2}^{m_1+1} \left\{ \prod_{j=0}^{i-2} [1 - \lambda(j)] \right\}$$
$$+ D_1 \sum_{i=2}^{n_2+1} \left\{ \prod_{j=0}^{i-2} [1 - \mu_1(j)] \right\} + D_2 \sum_{i=2}^{n_3+1} \left\{ \prod_{j=0}^{i-2} [1 - \mu_2(j)] \right\}.$$

Prove: According to Theorem 5.19, $\gamma = 1$ is the simple characteristic root of matrix B, which is equivalent to

$$\frac{d\,|B - \gamma I|}{d\gamma} \Big|_{\gamma=1} \neq 0. \tag{5.20}$$

Then, Formula 5.20 is proven:

$$\frac{d\,|B-\gamma I|}{d\gamma}\bigg|_{\gamma=1}$$

$$=\begin{bmatrix}
-1 & 0 & \cdots & 0 & 0 & \mu_1(0) & \mu_1(1) & \cdots \\
0 & -1 & \cdots & 0 & 0 & 0 & 0 & \cdots \\
0 & 1-\lambda(1) & \cdots & 0 & 0 & 0 & 0 & \cdots \\
\cdots & \cdots & \cdots & \cdots & \cdots & \cdots & \cdots & \cdots \\
0 & 0 & \cdots & 1-\lambda(n_1-1) & -1 & 0 & 0 & \cdots \\
0 & 0 & \cdots & 0 & 0 & -1 & 0 & \cdots \\
0 & 0 & \cdots & 0 & 0 & 1-\mu_1(0) & -1 & \cdots \\
0 & 0 & \cdots & 0 & 0 & 0 & 1-\mu_1(1) & \cdots \\
\cdots & \cdots & \cdots & \cdots & \cdots & \cdots & \cdots & \cdots \\
0 & 0 & \cdots & 0 & 0 & 0 & 0 & \cdots \\
0 & \lambda(1) & \cdots & \lambda(n_1-1) & 1 & 0 & 0 & \cdots \\
0 & 0 & \cdots & 0 & 0 & 0 & 0 & \cdots \\
0 & 0 & \cdots & 0 & 0 & 0 & 0 & \cdots \\
\cdots & \cdots & \cdots & \cdots & \cdots & \cdots & \cdots & \cdots \\
0 & 0 & \cdots & 0 & 0 & 0 & 0 & \cdots
\end{bmatrix}$$

$$\begin{bmatrix}
\mu_1(n_2-1) & 1 & 0 & 0 & \cdots & 0 & 0 \\
0 & 0 & 0 & 0 & \cdots & 0 & 0 \\
0 & 0 & 0 & 0 & \cdots & 0 & 0 \\
\cdots & \cdots & \cdots & \cdots & \cdots & \cdots & \cdots \\
0 & 0 & 0 & 0 & \cdots & 0 & 0 \\
0 & 0 & \mu_2(0) & \mu_2(1) & \cdots & \mu_2(n_3-1) & 1 \\
0 & 0 & 0 & 0 & \cdots & 0 & 0 \\
0 & 0 & 0 & 0 & \cdots & 0 & 0 \\
\cdots & \cdots & \cdots & \cdots & \cdots & \cdots & \cdots \\
1-\mu_1(n_2-1) & -1 & 0 & 0 & \cdots & 0 & 0 \\
0 & 0 & -1 & 0 & \cdots & 0 & 0 \\
0 & 0 & 1-\mu_2(0) & -1 & \cdots & 0 & 0 \\
0 & 0 & 0 & 1-\mu_2(1) & \cdots & 0 & 0 \\
\cdots & \cdots & \cdots & \cdots & \cdots & \cdots & \cdots \\
0 & 0 & 0 & 0 & \cdots & 1-\mu_2(n_3-1) & -1
\end{bmatrix}$$

$$
+\begin{bmatrix}
-1 & 0 & \cdots & 0 & 0 & \mu_1(0) & \mu_1(1) & \cdots \\
1-\lambda(0) & -1 & \cdots & 0 & 0 & 0 & 0 & \cdots \\
0 & 0 & \cdots & 0 & 0 & 0 & 0 & \cdots \\
\cdots & \cdots & & \cdots & \cdots & \cdots & \cdots & \cdots \\
0 & 0 & \cdots & 1-\lambda(n_1-1) & -1 & 0 & 0 & \cdots \\
0 & 0 & \cdots & 0 & 0 & -1 & 0 & \cdots \\
0 & 0 & \cdots & 0 & 0 & 1-\mu_1(0) & -1 & \cdots \\
0 & 0 & \cdots & 0 & 0 & 0 & 1-\mu_1(1) & \cdots \\
\cdots & \cdots & & \cdots & \cdots & \cdots & \cdots & \cdots \\
0 & 0 & \cdots & 0 & 0 & 0 & 0 & \cdots \\
\lambda(0) & 0 & \cdots & \lambda(n_1-1) & 1 & 0 & 0 & \cdots \\
0 & 0 & \cdots & 0 & 0 & 0 & 0 & \cdots \\
0 & 0 & \cdots & 0 & 0 & 0 & 0 & \cdots \\
\cdots & \cdots & \cdots & \cdots & \cdots & \cdots & \cdots & \cdots \\
0 & 0 & \cdots & 0 & 0 & 0 & 0 & \cdots
\end{bmatrix}
$$

$$
\begin{bmatrix}
\mu_1(n_2-1) & 1 & 0 & 0 & \cdots & 0 & 0 \\
0 & 0 & 0 & 0 & \cdots & 0 & 0 \\
0 & 0 & 0 & 0 & \cdots & 0 & 0 \\
\cdots & \cdots & \cdots & \cdots & \cdots & \cdots & \cdots \\
0 & 0 & 0 & 0 & \cdots & 0 & 0 \\
0 & 0 & \mu_2(0) & \mu_2(1) & \cdots & \mu_2(n_3-1) & 1 \\
0 & 0 & 0 & 0 & \cdots & 0 & 0 \\
0 & 0 & 0 & 0 & \cdots & 0 & 0 \\
\cdots & \cdots & \cdots & \cdots & \cdots & \cdots & \cdots \\
1-\mu_1(n_2-1) & -1 & 0 & 0 & \cdots & 0 & 0 \\
0 & 0 & -1 & 0 & \cdots & 0 & 0 \\
0 & 0 & 1-\mu_2(0) & -1 & \cdots & 0 & 0 \\
0 & 0 & 0 & 1-\mu_2(1) & \cdots & 0 & 0 \\
\cdots & \cdots & \cdots & \cdots & \cdots & \cdots & \cdots \\
0 & 0 & 0 & 0 & \cdots & 1-\mu_2(n_3-1) & -1
\end{bmatrix}+\cdots
$$

$$
+ \begin{bmatrix}
-1 & 0 & \cdots & 0 & 0 & \mu_1(0) & \mu_1(1) & \cdots \\
1-\lambda(0) & -1 & \cdots & 0 & 0 & 0 & 0 & \cdots \\
0 & 1-\lambda(1) & \cdots & 0 & 0 & 0 & 0 & \cdots \\
\cdots & \cdots & \cdots & \cdots & \cdots & \cdots & \cdots & \cdots \\
0 & 0 & \cdots & 1-\lambda(n_1-1) & -1 & 0 & 0 & \cdots \\
0 & 0 & \cdots & 0 & 0 & -1 & 0 & \cdots \\
0 & 0 & \cdots & 0 & 0 & 1-\mu_1(0) & -1 & \cdots \\
0 & 0 & \cdots & 0 & 0 & 0 & 1-\mu_1(1) & \cdots \\
\cdots & \cdots & \cdots & \cdots & \cdots & \cdots & \cdots & \cdots \\
0 & 0 & \cdots & 0 & 0 & 0 & 0 & \cdots \\
\lambda(0) & \lambda(1) & \cdots & \lambda(n_1-1) & 0 & 0 & 0 & \cdots \\
0 & 0 & \cdots & 0 & 0 & 0 & 0 & \cdots \\
0 & 0 & \cdots & 0 & 0 & 0 & 0 & \cdots \\
\cdots & \cdots & \cdots & \cdots & \cdots & \cdots & \cdots & \cdots \\
0 & 0 & \cdots & 0 & 0 & 0 & 0 & \cdots
\end{bmatrix}
$$

$$
\begin{bmatrix}
\mu_1(n_2-1) & 1 & 0 & 0 & \cdots & 0 & 0 \\
0 & 0 & 0 & 0 & \cdots & 0 & 0 \\
0 & 0 & 0 & 0 & \cdots & 0 & 0 \\
\cdots & \cdots & \cdots & \cdots & \cdots & \cdots & \cdots \\
0 & 0 & 0 & 0 & \cdots & 0 & 0 \\
0 & 0 & \mu_2(0) & \mu_2(1) & \cdots & \mu_2(n_3-1) & 1 \\
0 & 0 & 0 & 0 & \cdots & 0 & 0 \\
0 & 0 & 0 & 0 & \cdots & 0 & 0 \\
\cdots & \cdots & \cdots & \cdots & \cdots & \cdots & \cdots \\
1-\mu_1(n_2-1) & -1 & 0 & 0 & \cdots & 0 & 0 \\
0 & 0 & -1 & 0 & \cdots & 0 & 0 \\
0 & 0 & 1-\mu_2(0) & -1 & \cdots & 0 & 0 \\
0 & 0 & 0 & 1-\mu_2(1) & \cdots & 0 & 0 \\
\cdots & \cdots & \cdots & \cdots & \cdots & \cdots & \cdots \\
0 & 0 & 0 & 0 & \cdots & 1-\mu_2(n_3-1) & -1
\end{bmatrix} + \cdots
$$

$$
+ \begin{bmatrix}
-1 & 0 & \cdots & 0 & 0 & \mu_1(0) & \mu_1(1) & \cdots \\
1-\lambda(0) & -1 & \cdots & 0 & 0 & 0 & 0 & \cdots \\
0 & 1-\lambda(1) & \cdots & 0 & 0 & 0 & 0 & \cdots \\
\cdots & \cdots & \cdots & \cdots & \cdots & \cdots & \cdots & \cdots \\
0 & 0 & \cdots & 1-\lambda(n_1-1) & -1 & 0 & 0 & \cdots \\
0 & 0 & \cdots & 0 & 0 & -1 & 0 & \cdots \\
0 & 0 & \cdots & 0 & 0 & 1-\mu_1(0) & -1 & \cdots \\
0 & 0 & \cdots & 0 & 0 & 0 & 1-\mu_1(1) & \cdots \\
\cdots & \cdots & \cdots & \cdots & \cdots & \cdots & \cdots & \cdots \\
0 & 0 & \cdots & 0 & 0 & 0 & 0 & \cdots \\
\lambda(0) & \lambda(1) & \cdots & \lambda(n_1-1) & 1 & 0 & 0 & \cdots \\
0 & 0 & \cdots & 0 & 0 & 0 & 0 & \cdots \\
0 & 0 & \cdots & 0 & 0 & 0 & 0 & \cdots \\
\cdots & \cdots & \cdots & \cdots & \cdots & \cdots & \cdots & \cdots \\
0 & 0 & \cdots & 0 & 0 & 0 & 0 & \cdots
\end{bmatrix}
$$

$$
\begin{bmatrix}
\mu_1(n_2) & 1 & 0 & 0 & \cdots & 0 & 0 \\
0 & 0 & 0 & 0 & \cdots & 0 & 0 \\
0 & 0 & 0 & 0 & \cdots & 0 & 0 \\
\cdots & \cdots & \cdots & \cdots & \cdots & \cdots & \cdots \\
0 & 0 & 0 & 0 & \cdots & 0 & 0 \\
0 & 0 & \mu_2(0) & \mu_2(1) & \cdots & \mu_2(n_3-1) & 0 \\
0 & 0 & 0 & 0 & \cdots & 0 & 0 \\
0 & 0 & 0 & 0 & \cdots & 0 & 0 \\
\cdots & \cdots & \cdots & \cdots & \cdots & \cdots & \cdots \\
1-\mu_1(n_2-1) & -1 & 0 & 0 & \cdots & 0 & 0 \\
0 & 0 & -1 & 0 & \cdots & 0 & 0 \\
0 & 0 & 1-\mu_2(0) & -1 & \cdots & 0 & 0 \\
0 & 0 & 0 & 1-\mu_2(1) & \cdots & 0 & 0 \\
\cdots & \cdots & \cdots & \cdots & \cdots & \cdots & \cdots \\
0 & 0 & 0 & 0 & \cdots & 1-\mu_2(n_3-1) & -1
\end{bmatrix}
$$

$$
= (-1)^{n_1+n_2+n_3+2}\{-2+(\lambda(0)-1]+\ldots+[\lambda(n_1-1)-1]+[\mu_1(0)-1]+\ldots
$$

$$
+\ [\mu_1(n_2-1)-1]+[\lambda(n_1)-1]+[\mu_2(0)-1]+\ldots+[\mu_2(n_3-1)-1]\}\neq 0.
$$

Thus, $\gamma = 1$ is the simple characteristic root of matrix B.

Suppose X_0 stands for the feature vector of $\gamma = 1$ on C. Obviously, $(B-I)X_0 = 0$; that is,

$$
\left[
\begin{array}{ccccccccccccccc}
-1 & 0 & \cdots & 0 & 0 & \mu_1(0) & \mu_1(1) & \cdots & \mu_1(n_2-1) & 1 & \mu_2(0) & \mu_2(1) & \cdots & \mu_2(n_3-1) & 1 \\
1-\lambda(0) & -1 & \cdots & 0 & 0 & 0 & 0 & \cdots & 0 & 0 & 0 & 0 & \cdots & 0 & 0 \\
0 & 1-\lambda(1) & \cdots & 0 & 0 & 0 & 0 & \cdots & 0 & 0 & 0 & 0 & \cdots & 0 & 0 \\
\cdots & \cdots & \cdots & \cdots & \cdots & \cdots & \cdots & \cdots & \cdots & \cdots & \cdots & \cdots & \cdots & \cdots & \cdots \\
0 & 0 & \cdots & 1-\lambda(n_1-1) & -1 & 0 & 0 & \cdots & 0 & 0 & 0 & 0 & \cdots & 0 & 0 \\
\lambda(0) & \lambda(1) & \cdots & \lambda(n_1-1) & \lambda(n_1) & -1 & 0 & \cdots & 0 & 0 & 0 & 0 & \cdots & 0 & 0 \\
0 & 0 & \cdots & 0 & 0 & 1-\mu_1(0) & -1 & \cdots & 0 & 0 & 0 & 0 & \cdots & 0 & 0 \\
0 & 0 & \cdots & 0 & 0 & 0 & 1-\mu_1(1) & \cdots & 0 & 0 & 0 & 0 & \cdots & 0 & 0 \\
\cdots & \cdots & \cdots & \cdots & \cdots & \cdots & \cdots & \cdots & \cdots & \cdots & \cdots & \cdots & \cdots & \cdots & \cdots \\
0 & 0 & \cdots & 0 & 0 & 0 & 0 & \cdots & 1-\mu_1(n_2-1) & -1 & 0 & 0 & \cdots & 0 & 0 \\
0 & 0 & \cdots & 0 & 1-\lambda(n_1) & 0 & 0 & \cdots & 0 & 0 & -1 & 0 & \cdots & 0 & 0 \\
0 & 0 & \cdots & 0 & 0 & 0 & 0 & \cdots & 0 & 0 & 1-\mu_2(0) & -1 & \cdots & 0 & 0 \\
0 & 0 & \cdots & 0 & 0 & 0 & 0 & \cdots & 0 & 0 & 0 & 1-\mu_2(1) & \cdots & 0 & 0 \\
\cdots & \cdots & \cdots & \cdots & \cdots & \cdots & \cdots & \cdots & \cdots & \cdots & \cdots & \cdots & \cdots & \cdots & \cdots \\
0 & 0 & \cdots & 0 & 0 & 0 & 0 & \cdots & 0 & 0 & 0 & 0 & \cdots & 1-\mu_2(n_3-1) & -1
\end{array}
\right] X_0 = 0
$$

According to Theorem 5.20, the rank of matrix $I-B$ is $n_1 + n_2 + n_3 + 2$. Then,

$$
\begin{cases}
[1-\lambda(0)]x_1 - x_2 = 0 \\[4pt]
[1-\lambda(1)]x_2 - x_3 = 0 \\[4pt]
\cdots \\[4pt]
[1-\lambda(n_1-1)]x_{n_1} - x_{n_1+1} = 0 \\[4pt]
\displaystyle\sum_{i=1}^{n_1+1}[x_i\lambda(i-1)] - x_{n_1+2} = 0 \\[4pt]
[1-\mu_2(0)]x_{n_1+2} - x_{n_1+3} = 0 \\[4pt]
[1-\mu_2(1)]x_{n_1+3} - x_{n_1+4} = 0 \\[4pt]
\cdots \\[4pt]
[1-\mu_1(n_2-1)]x_{n_1+n_2+1} - x_{n_1+n_2+2} = 0 \\[4pt]
x_{n_1+1}[1-\lambda(n_1)] - x_{n_1+n_2+3} = 0 \\[4pt]
[1-\mu_2(0)]x_{n_1+n_2+3} - x_{n_1+n_2+4} = 0 \\[4pt]
[1-\mu_2(1)]x_{n_1+n_2+4} - x_{n_1+n_2+5} = 0 \\[4pt]
\cdots \\[4pt]
[1-\mu_2(n_2-1)]x_{n_1+n_2+n_3+2} - x_{n_1+n_2+n_3+3} = 0.
\end{cases}
$$

In other words,

$$
x_i = \prod_{j=0}^{i-2}[1-\lambda(j)]\,x_1, \quad i = 2,3,\ldots, n_1+1
$$

$$
x_{n_1+i+1} = \prod_{j=0}^{i-2}[1-\mu_1(j)]\,x_{n_1+2}, \quad i = 2,3,\ldots, n_2+1
$$

$$
x_{n_1+n_2+i+2} = \prod_{j=0}^{i-2}[1-\mu_2(j)]\,x_{n_1+n_2+3}, \quad i = 2,3,\ldots, n_3+1
$$

$$\sum_{i=1}^{n_1+1} [x_i \lambda(i-1)] - x_{m+2} = 0$$

$$x_{m+1}[1 - \lambda(n_1)] - x_{m+n_2+3} = 0.$$

Suppose

$$D_1 = \lambda(0) + \sum_{i=2}^{n_1+1} \left\{ \prod_{j=0}^{i-2} [1 - \lambda(j)] \lambda(i-1) \right\}$$

$$D_2 = \prod_{j=0}^{n_1-1} [1 - \lambda(j)][1 - \lambda(n_1)].$$

Then,

$$x_{m+2} = D_1 x_1$$

$$x_{m+n_2+3} = D_2 x_1.$$

The following conclusion is easily proven:

$$D_1 + D_2 = 1.$$

Because $X_0 \in C$; that is,

$$\left(1 + \sum_{i=2}^{n_1+1} \left\{ \prod_{j=0}^{i-2} [1 - \lambda(j)] \right\} \right) x_1 + \left(1 + \sum_{i=2}^{n_2+1} \left\{ \prod_{j=0}^{i-2} [1 - \mu_1(j)] \right\} \right) D_1 x_1$$

$$+ \left(1 + \sum_{i=2}^{n_3+1} \left\{ \prod_{j=0}^{i-2} [1 - \mu_2(j)] \right\} \right) D_2 x_1 = 1.$$

In other words,

$$\left(2 + \sum_{i=2}^{n_1+1} \left\{ \prod_{j=0}^{i-2} [1 - \lambda(j)] \right\} \right) x_1 + D_1 x_1 + \sum_{i=2}^{n_2+1} \left\{ \prod_{j=0}^{i-2} [1 - \mu(j)] \right\}$$

$$+ D_2 x_1 + \sum_{i=2}^{n_3+1} \left\{ \prod_{j=0}^{i-2} [1 - \rho(j)] \right\} = 1.$$

Suppose

$$D = 2 + \sum_{i=2}^{n_1+1}\left\{\prod_{j=0}^{i-2}[1-\lambda(j)]\right\} + D_1\sum_{i=2}^{n_2+1}\left\{\prod_{j=0}^{i-2}[1-\mu(j)]\right\}$$

$$+ D_2\sum_{i=2}^{n_3+1}\left\{\prod_{j=0}^{i-2}[1-\rho(j)]\right\}.$$

Then,

$$x_1 = \frac{1}{D}$$

$$x_i = \frac{1}{D}\prod_{j=0}^{i-2}[1-\lambda(j)], \quad i = 2,3,\ldots, n_1 + 1$$

$$x_{n_1+2} = \frac{D_1}{D}$$

$$x_{n_1+i+1} = \frac{D_1}{D}\prod_{j=0}^{i-2}[1-\mu_1(j)], \quad i = 2,3,\ldots, n_2 + 1$$

$$x_{n_1+n_2+3} = \frac{D_2}{D}$$

$$x_{n_1+n_2+i+1} = \frac{D_2}{D}\prod_{j=0}^{i-2}[1-\mu_2(j)], \quad i = 2,3,\ldots, n_3 + 1.$$

Theorem 5.22: The characteristic root of matrix B that corresponds to $\rho(B) = 1$ can only be $\gamma = 1$.

Prove: Suppose that the characteristic root of matrix B that corresponds to $\rho(B) = 1$ is $\gamma = e^{bw}$, in which $\gamma = e^{bw}$ and $w^2 = -1$. Suppose the feature vector corresponding to characteristic root $\gamma = e^{bw}$ of matrix B is $\zeta = (x_1, x_2, \ldots, x_{n_1+n_2+n_3+3})^T \neq 0$.

Then,

$$(B - \gamma I)\zeta = 0.$$

In other words,

$$
\left\{
\begin{aligned}
&-\gamma x_1 + \sum_{i=1}^{n_2+1}[x_{m_1+i+1}\mu_1(i-1)] + \sum_{i=1}^{n_3+1}[x_{m_1+n_2+i+2}\mu_2(i-1)] = 0 \\
&[1-\lambda(0)]x_1 - \gamma x_2 = 0 \\
&[1-\lambda(1)]x_2 - \gamma x_3 = 0 \\
&\cdots \\
&[1-\lambda(n_1-1)]x_{m_1} - \gamma x_{m_1+1} = 0 \\
&\sum_{i=1}^{n_3+1}[x_i\lambda(i-1)] - \gamma x_{m_1+2} = 0 \\
&[1-\mu_2(0)]x_{m_1+2} - \gamma x_{m_1+3} = 0 \\
&[1-\mu_2(1)]x_{m_1+3} - \gamma x_{m_1+4} = 0 \\
&\cdots \\
&[1-\mu_1(n_2-1)]x_{m_1+n_2+1} - \gamma x_{m_1+n_2+2} = 0 \\
&x_{m_1+1}[1-\lambda(n_1)] - \gamma x_{m_1+n_2+3} = 0 \\
&[1-\mu_2(0)]x_{m_1+n_2+3} - \gamma x_{m_1+n_2+4} = 0 \\
&[1-\mu_2(1)]x_{m_1+n_2+4} - \gamma x_{m_1+n_2+5} = 0 \\
&\cdots \\
&[1-\mu_2(n_2-1)]x_{m_1+n_2+n_3+2} - \gamma x_{m_1+n_2+n_3+3} = 0.
\end{aligned}
\right.
$$

Then,

$$
x_i = \prod_{j=0}^{i-2}\left[\frac{1-\lambda(j)}{\gamma}\right]x_1, \quad i = 2,3,\ldots,n_1+1
$$

$$
x_{m_1+i+1} = \prod_{j=0}^{i-2}\left[\frac{1-\mu_1(j)}{\gamma}\right]x_{m_1+2}, \quad i = 2,3,\ldots,n_2+1
$$

$$
x_{m_1+n_2+i+2} = \prod_{j=0}^{i-2}\left[\frac{1-\mu_2(j)}{\gamma}\right]x_{m_1+n_2+3}, \quad i = 2,3,\ldots,n_3+1
$$

$$
\sum_{i=1}^{n_1+1}[x_i\lambda(i-1)] - \frac{1}{\gamma}x_{m_1+2} = 0
$$

$$x_{m_1+1}[1-\lambda(n_1)]-\frac{1}{\gamma}x_{m_1+n_2+3}=0$$

$$-x_1+\sum_{i=1}^{n_2+1}\left[x_{m_1+i+1}\frac{\mu_1(i-1)}{\gamma}\right]+\sum_{i=1}^{n_3+1}\left[x_{m_1+n_2+i+2}\frac{\mu_2(i-1)}{\gamma}\right]=0.$$

Suppose

$$D_1=\lambda(0)e^{-bw}+\sum_{i=2}^{n_1+1}\left\{\prod_{j=0}^{i-2}[1-\lambda(j)]\lambda(i-1)e^{-biw}\right\}$$

$$D_2=\sum_{j=0}^{m_1-1}[1-\lambda(j)][1-\lambda(n_1)]e^{-b(m_1+1)w}$$

$$D_3=\mu_1(0)e^{-bw}+\sum_{i=2}^{n_2+1}\left\{\prod_{j=0}^{i-2}[1-\mu_1(j)]\mu_1(i-1)e^{-biw}\right\}$$

$$D_4=\mu_2(0)e^{-bw}+\sum_{i=2}^{n_2+1}\left\{\prod_{j=0}^{i-2}[1-\mu_2(j)]\mu_2(i-1)e^{-biw}\right\}.$$

Then,

$$x_1=D_3x_{m_1+2}+D_4x_{m_1+n_2+3}$$

$$x_{m_1+2}=D_1x_1$$

$$x_{m_1+n_2+3}=D_2x_1.$$

Thus,

$$(D_3D_1+D_4D_2-1)x_1=0.$$

Obviously, $x_1\neq 0$. Otherwise, $x_i=0$ and $i=1, 2, \ldots, n_1+n_2+n_3+3$, which is contradictory with $\zeta=\left(x_1,x_2,\ldots,x_{2n+2}\right)^T\neq 0$. Then,

$$D_3D_1+D_4D_2=1$$

$$|D_3 D_1 + D_4 D_2| = 1.$$

Because

$$|D_1| = \left| \lambda(0)e^{-bw} + \sum_{i=2}^{n_1+1} \left\{ \prod_{j=0}^{i-2} [1-\lambda(j)]\lambda(i-1)e^{-biw} \right\} \right|$$

$$\leq \left| \lambda(0)e^{-bw} \right| + \sum_{i=2}^{n_1+1} \left\{ \prod_{j=0}^{i-2} [1-\lambda(j)]\lambda(i-1) \left| e^{-biw} \right| \right\}$$

$$= \lambda(0) + \sum_{i=2}^{n_1+1} \left\{ \prod_{j=0}^{i-2} [1-\lambda(j)]\lambda(i-1) \right\}$$

$$|D_2| = \left| \prod_{j=0}^{n_1-1} [1-\lambda(j)][1-\lambda(n_1)e^{-b(n_1+1)w}] \right| = \prod_{j=0}^{n_1-1} [1-\lambda(j)][1-\lambda(n_1)]$$

$$|D_3| = \left| \mu_1(0)e^{-bw} + \sum_{i=2}^{n_2+1} \left\{ \prod_{j=0}^{i-2} [1-\mu_1(j)]\mu_1(i-1)e^{-biw} \right\} \right|$$

$$\leq \mu_1(0) \left| e^{-bw} \right| + \sum_{i=2}^{n_2+1} \left\{ \prod_{j=0}^{i-2} [1-\mu_1(j)]\mu_1(i-1) \left| e^{-biw} \right| \right\}$$

$$= \mu_1(0) + \sum_{i=2}^{n_2+1} \left\{ \prod_{j=0}^{i-2} [1-\mu_1(j)]\mu_1(i-1) \right\} = 1$$

$$|D_4| = \left| \mu_2(0)e^{-bw} + \sum_{i=2}^{n_3+1} \left\{ \prod_{j=0}^{i-2} [1-\mu_2(j)]\mu_2(i-1)e^{-biw} \right\} \right|$$

$$\leq \mu_2(0) \left| e^{-bw} \right| + \sum_{i=2}^{n_3+1} \left\{ \prod_{j=0}^{i-2} [1-\mu_2(j)]\mu_2(i-1) \left| e^{-biw} \right| \right\}$$

$$= \mu_2(0) + \sum_{i=2}^{n_3+1} \left\{ \prod_{j=0}^{i-2} [1-\mu_2(j)]\mu_2(i-1) \right\} = 1,$$

the following conclusion is easily proven:

$$|D_1| + |D_2| = 1.$$

According to Reference [84], the necessary and sufficient conditions for establishment of an equal sign in $|D_3\|D_1| + |D_4\|D_2| \leq 1$ are as follows:

1. All vectors expressed by $\lambda(0)e^{-bw}$ and $\prod_{j=0}^{i-2}[1-\lambda(j)]\lambda(i-1)e^{-biw}$ $(i = 2, 3, \ldots,$

   $n_1 + 1)$ are collinear with the same direction.

2. All vectors expressed by $\mu_1(0)e^{-bw}$ and $\prod_{j=0}^{i-2}[1-\mu_1(j)]\mu_1(i-1)e^{-biw}$ $(i = 2, 3, \ldots,$

   $n_2 + 1)$ are collinear with the same direction.

3. All vectors expressed by $\mu_2(0)e^{-bw}$ and $\prod_{j=0}^{i-2}[1-\mu_2(j)]\mu_2(i-1)e^{-biw}$ $(i = 2, 3, \ldots,$

   $n_3 + 1)$ are collinear with the same direction.

According to the value range of $0 \leq b < 2\pi$, $b = 0$ can be obtained. Thus, the characteristic root of matrix B corresponding to $\rho(B) = 1$ can only be $\gamma = 1$.

The following conclusions can be obtained according to Theorems 5.19–5.22.

Theorem 5.23: Status solution (5.19) of systems (5.15 through 5.17) is stable and the unique equilibrium point is X_0; that is, feature vector of characteristic root $\gamma = 1$ of matrix B on C.

Prove: According to Theorem 5.8.6 in Reference [85], the status solution (5.19) of systems (5.15 through 5.17) is stable. Suppose the equilibrium point is X_1, $X_1 \in C$ and $X_1 = BX_1$. Then, X_1 is a feature vector of characteristic root $\gamma = 1$ of matrix B on C. Then, $X_1 = X_0$ can be obtained as per Theorem 5.21.

Theorem 5.24: System instantaneous availability (5.17) is stable, and the equilibrium point is unique; that is, the steady-state availability of system is unique with value as follows:

$$A = \sum_{i=1}^{n_1+1} x_i = \frac{1}{D}\left(1 + \sum_{i=2}^{n_1+1}\left\{\prod_{j=0}^{i-2}[1-\lambda(j)]\right\}\right). \tag{5.21}$$

In the formula,

$$D = 2 + \sum_{i=2}^{n_1+1}\left\{\prod_{j=0}^{i-2}[1-\lambda(j)]\right\} + D_1\sum_{i=2}^{n_2+1}\left\{\prod_{j=0}^{i-2}[1-\mu_1(j)]\right\} + D_2\sum_{i=2}^{n_3+1}\left\{\prod_{j=0}^{i-2}[1-\mu_2(j)]\right\}$$

$$D_1 = \lambda(0) + \sum_{i=2}^{n_1+1}\left\{\prod_{j=0}^{i-2}[1-\lambda(j)]\lambda(i-1)\right\}$$

$$D_2 = \prod_{j=0}^{n_1-1}[1-\lambda(j)][1-\lambda(n_1)].$$

Prove: According to Theorem 5.23, the following formula can be obtained:

$$X_0 = \lim_{k\to\infty} P(k).$$

According to Theorem 5.21, the limit

$$A = \lim_{k\to\infty} A(k) = \lim_{k\to\infty} \sigma_{n_1+1,n_2+n_3+2}P(k) = \sigma_{n_1+1,n_2+n_3+2}X_0$$

$$= \sum_{i=1}^{n_1+1} x_i = \frac{1}{D}\left(1 + \sum_{i=2}^{n_1+1}\left\{\prod_{j=0}^{i-2}[1-\lambda(j)]\right\}\right)$$

exists; that is, the steady-state availability of a system.

5.5 Repairable System with Complex Structure

5.5.1 System Instantaneous Availability Model

Suppose the system is composed of n units; both unit and system only have two kinds of status (i.e., normal and failed), which are expressed with 0 and 1, respectively. The status of all units is mutually independent. If system structure function is $\phi(.)$ and reliability function [8] of structure ϕ is $h(.)$, the instantaneous availability of system is

$$A(t) = h(A_1(t), A_2(t), \ldots, A_n(t)). \tag{5.22}$$

In the formula, $A_i(t)$ stands for the instantaneous availability function of unit i and $i = 1, 2, \ldots, n$, which can satisfy one of instantaneous availability models of all systems in Sections 5.2–5.5 of this chapter, according to actual situations.

5.5.2 *Stability Proving of Instantaneous Availability*

Theorem 5.25: System instantaneous availability is stable, and the equilibrium point is unique; that is, the steady-state availability of system is unique with a value as follows:

$$A = h(A_1, A_2, \ldots, A_n). \tag{5.23}$$

In the formula, A_i stands for steady-state availability of unit i.

Prove: From previous paragraphs, it can be seen that the instantaneous availability of all units is stable. In other words, the steady-state availability exists; that is,

$$A_i = \lim_{t \to \infty} A_i(t), \quad i = 1, 2, \ldots, n.$$

Then, according to features of structure function and formula (5.22),

$$\lim_{t \to \infty} A(t) = \lim_{t \to \infty} h(A_1(t), A_2(t), \ldots, A_n(t))$$

$$= h\left(\lim_{t \to \infty} A_1(t), \lim_{t \to \infty} A_2(t), \ldots, \lim_{t \to \infty} A_n(t)\right)$$

$$= h(A_1, A_2, \ldots, A_n).$$

5.6 Summary

In Sections 5.2–5.4 of this chapter, instantaneous availability models of various single-unit repairable systems with finite time restrained are established through which the stability of instantaneous availability is proven and an expression of steady-state availability is obtained. Through research on a complex repairable system in Section 5.5, it is found that if the structure of a complex repairable system is given, the research on instantaneous availability of such a complex repairable system can be decomposed to study several single-unit repairable systems. Therefore, in this chapter, the research is mainly carried out on instantaneous availability of single-unit repairable systems. In other words, the research on complex structure was performed based on instantaneous availability models in Sections 5.2–5.4.

Chapter 6

Analysis and Control Method for Volatility of System Instantaneous Availability

6.1 Introduction

Instantaneous availability models of typical repairable systems with infinite time restrained are established in Chapter 4. Then, in Chapter 5, it was proven that the instantaneous availability $A(k)$ of such systems are all gradually convergent to steady-state availability A; that is, $A(k) \xrightarrow{k \to \infty} A$, which indicates that when the time (t) reaches a certain moment, the instantaneous availability will stably stay in a very small range of nearby steady-state availability. However, the convergency cannot ensure that the instantaneous availability has no large volatility during a finite period. To illustrate this, two systems are compared in this chapter. The service life and repair time of system 1 are both subject to discrete Weibull distribution with a shape parameter of 1 and a dimension parameter of 0.98, which degrades into geometric distribution; the service life and repair time of system 2 are both subject to discrete Weibull distribution with a shape parameter of 3 and a dimension parameter of 0.99998. The comparison result of instantaneous availability curves of these two systems is shown in Figure 6.1, where it can be

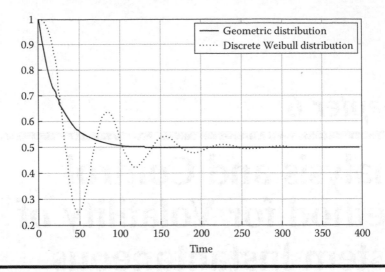

Figure 6.1 Comparison of instantaneous availabilities of two different systems with the same inherent availability.

seen that, although the inherent availability or steady-state availability of the two systems is the same, the volatility of their instantaneous availability during the initial period is largely different. Then how can the volatility characteristics of system instantaneous availability be described? And what does this have to do with service life, repair time, or other related times of systems?

Through contrastive analysis of the two systems in Figure 6.1, it appears that similar volatilities of system instantaneous availability are caused by the incompatibility of the maintenance subsystem and the support subsystem of the equipment system with the equipment itself. When the system instantaneous availability fluctuates wildly, this indicates that the probability of the equipment functioning normally during some period is relatively low. It may be that the equipment has not formed its due efficiency at this moment and that further consistency is required for subsystems (i.e., the new equipment shall perform effectively). Therefore, it is very significant to research characteristics of the course before the instantaneous availability of the equipment reaches steady-state availability. The research focus of this chapter is on coordinating subsystems to make system instantaneous availability meet specified requirements as soon as possible and to effectively overcome or reduce such volatility of instantaneous availability.

In this chapter, the matching problems of new equipment systems that are caused by interaction of subsystems are researched through an analysis of volatility (variation) of instantaneous availability during a finite period. Based on research needs, some concepts are proposed and future research content and framework are pointed out in this chapter.

6.2 Volatility Parameters of System Instantaneous Availability

In order to better research the volatility characteristic of system instantaneous availability and find out its inherent law, a parameter system will be presented first to describe the volatility characteristic of system instantaneous availability.

6.2.1 Indicative Function

Definition 6.1: Suppose $\varepsilon_0 > 0$, the function

$$j(k, \varepsilon_0) = \begin{cases} A - \varepsilon_0 - A(k) & \text{When } A(k) < A - \varepsilon_0 \\ 0 & \text{Others} \end{cases}, \quad k = 0, 1, 2, \ldots \quad (6.1)$$

is called the indicative function of a system in which A stands for the steady-state availability of the system. In addition,

$$J(k, \varepsilon_0) = \sum_{\tau = k}^{\infty} j(\tau, \varepsilon_0), \quad t \geq 0. \quad (6.2)$$

is called the indicative metric function and $\varepsilon_0 > 0$ is the availability calibration level.

Note 1: The availability calibration level ε_0 is a given index, which usually depends on the actual condition of a system and reflects the basic requirement for system instantaneous availability. The physical meaning of indicative function reflects the deviation between system instantaneous availability and critical value. Then, the indicative metric function is the total deviation after $k \geq 0$, which stands for the cumulative deviation (difference) between system instantaneous availability and critical value.

For the two systems presented as examples in this chapter's introduction, suppose an availability calibration level $\varepsilon_0 = 0.02$. In this case, the instantaneous availability of system 1 monotonically decreases and stabilizes to steady-state availability, and the system indicative function is $j(k, \varepsilon_0) \equiv 0$, $J(k, \varepsilon_0) \equiv 0$, $k = 0, 1, 2, \ldots$, which indicates that the system instantaneous availability has no volatility. The indicative function of system 2 is shown in Figure 6.2. From the figure, it can be seen that, near $\tau = 50, 118, 183, 262, 335$ unit time, the system indicative function reaches its extreme value and has a declining trend overall. Therefore, it appears that indicative function can well reflect the volatility degree of system instantaneous availability relative to critical value. The indicative metric function is shown in Figure 6.3, which reflects the cumulative volatility of system instantaneous availability.

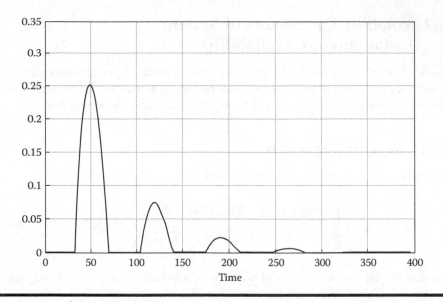

Figure 6.2   Indicative function of system 2.

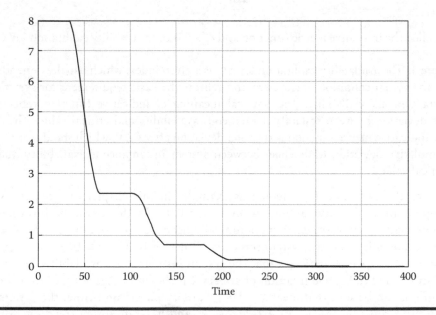

Figure 6.3   Indicative metric function of system 2.

For simplicity, the availability calibration level ε_0 is omitted in the following mathematical expressions of system indicative function and indicative metric function, which are called $j(k)$ and $J(k)$, respectively, for short. Obviously, $J(k)$ is featured as follows:

1. Nonnegativity: $J(t) \geq 0$
2. Monotonicity: $J(k_1) \leq J(k_2)$, when $k_1 \geq k_2$
3. Convergence: $\exists T_0, .s.t. J(k) = 0, \forall k \geq T_0$

6.2.2 System Matching and System Adaptation Time

Based on definitions of indicative function and indicative metric function, the system matching is defined through indicative function.

Definition 6.2: Suppose an availability calibration level ε_0 of the system is given. When its indicative function $j(k) = 0$ with $\forall k \in N$ (i.e., the system indicative metric function $J(0) = 0$), the system is matched; otherwise, the system is not matched.

Note 2: Matching is relative. Different results may be obtained for a different calibration level ε_0. To facilitate this description, suppose ε_0 has been given in this chapter.

Note 3: The physical meaning of a matched system indicates that the maintenance subsystem and support subsystem can coordinate with the equipment itself to make system instantaneous availability always meet specified requirements. In this chapter, the specified requirement refers to the fact that the instantaneous availability stays above a certain level.

For an unmatched system, the system adaptation time is defined in this chapter as describing the unmatching degree of an unmatched system.

Definition 6.3: System adaptation time T is defined as: $\forall k \geq T, J(k) = 0$; $\forall \zeta > 0$, and $J(T - \zeta) > 0$; $[0, T]$ is called a transitional period of system matching. When $J(0) = 0$, the supplementary definition $T = 0$ is established.

Note 4: According to the definition of system adaptation time, the indicative metric function still satisfies $J(k) = \sum_{\tau=k}^{T} j(\tau)$. In other words, when the availability calibration level is given, the system availability reaches a persistent relative stable status from volatility status from the earliest time of T. The system adaptation time to some extent reflects the convergence time from instantaneous availability to steady-state availability. According to the definition of adaptation time, the matched system can also be understood as a system with adaptation time = 0.

We continued to carry out analysis using the examples given in this chapter's introduction. Suppose the parameter $\varepsilon_0 = 0.02$. According to Definition 6.2 and to previous conclusions, system 1 is matched with adaptation time = 0. Therefore, it needs no matching transition; system 2 is unmatched with adaptation time $T = 200$. Its transitional period for matching is $[0, 200]$.

6.2.3 Availability Amplitude and Occurrence Time of Minimum Availability

For an unmatched system, a large amount of numerical simulation indicates that the volatility of system instantaneous availability usually decreases over time and trends to zero. However, in the initial period, the volatility degree of availability is different. In order to better describe the volatility characteristics of a system, two following concepts are given.

Definition 6.4: Definition $A_{\min} = \min_{k}\{A(k)\}$ is the minimum availability of the system, which occurs at the time of

$$T_0 = \min_{k}\{k \mid A(k) = A_{\min}\} > 0.$$

When the minimum value of $A(t)$ does not occur within a finite time, $T_0 = N$.

Definition 6.5: The maximum availability amplitude is defined as the maximum difference between instantaneous value (smaller) and inherent availability (larger):

$$M = A - A_{\min} \geq 0.$$

In this formula, symbol A stands for the steady-state availability of a system. To avoid confusion, it is known as the availability amplitude for short.

Note 5: Availability amplitude indicates the difference between minimum availability and inherent availability. The larger the availability amplitude is, the more severe the system availability volatility. The occurrence time of minimum availability is just the starting time of the minimum value of system availability—or the time when the maximum availability amplitude appears. If the system steady-state availability is given, the research on system availability amplitude may be equivalent to that of system minimum availability.

6.2.4 Interval Availability

A single-unit repairable system with preventive maintenance is used as an example here. Suppose the service life, corrective maintenance time, and preventive maintenance time of a unit are subject to probability distributions Weibull(α_1, β_1), Weibull(α_2, β_2), and Weibull(α_3, β_3), respectively, in which, $\alpha_1 = 0.99998$, $\beta_1 = 3$, $\alpha_2 = 0.999$, $\beta_2 = 2$, $\alpha_3 = 0.995$, and $\beta_3 = 3$. In Figure 6.4, instantaneous availability curves for a repairable system with a preventive maintenance cycle = 30 unit time, a repairable system with a preventive maintenance cycle = 60 unit time, and a repairable system without preventive maintenance are presented. From the figure, it can be seen that the preventive maintenance results in more severe volatility of system instantaneous availability, but it is helpful to the improvement of system steady-state availability. Therefore, the stationary factor of availability is sacrificed to improve the steady-state availability of the system.

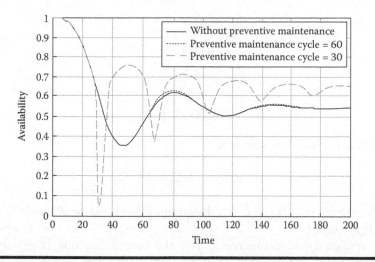

Figure 6.4 Curves for instantaneous availability of systems with different preventive maintenance cycles.

For this chapter, the volatility of instantaneous availability has been researched based on the same inherent steady-state availability. Obviously, preventive maintenance can improve the steady-state availability of a system. Therefore, it is meaningless to merely research instantaneous availability with no consideration of steady-state availability. Because the instantaneous availability of a system at the initial stage has been researched, it is already known that the steady-state availability is the convergence value of system instantaneous availability at infinity of time. Therefore, in this chapter, the average system availability at the initial operating stage of new equipment is used as a characteristic parameter to describe the volatility of system instantaneous availability, which can well describe the overall condition of system instantaneous availability at the initial operation stage of new equipment.

Because the previous analytical expression of instantaneous availability is unobtainable for research on interval availability, we can research the Markov repairable system or carry out research only on the estimated value obtained from statistics. Therefore, in this chapter, the interval availability of system is first obtained using the instantaneous availability model in Reference [86]; then, research is conducted.

6.2.5 Parameter of Stable Volatility Speed

6.2.5.1 Parameter Stability Concept

Suppose the system dynamic equation is

$$x(k + 1) = f(x(k), u(k)). \tag{6.3}$$

Output is

$$y(k) = g(x(k)). \tag{6.4}$$

Parameters are

$$r(k) = h(x(k)) \in R^l. \tag{6.5}$$

The initial value of the system is

$$x(0) = x_0 \in \Omega \subset R^n, \tag{6.6}$$

where the system state is $x(k) \in R^n$, the system input is $u(k) \in R^l$, and the system output is $y(k) \in R^m$.

The system dynamic function is $f(\cdot)$, the output function is $g(\cdot)$, and the parameters-system input/output relation function $h(\cdot)$ and system input/output block diagram are shown in Figure 6.5.

As for the steady speed of the parameter $r(k) \in R^l$, the problem to be solved first is the one about parameter $r(k)$. Taking into account the complexity of the problem, it should be started first using the linear system.

Taking into account that the system

$$x(k+1) = Ax(k) + b, \tag{6.7}$$

output is

$$y(k) = Cx(k) + d. \tag{6.8}$$

The parameter is

$$r(k) = Px(k) + e. \tag{6.9}$$

The initial state is

$$x(0) = x_0 \in \Omega, \tag{6.10}$$

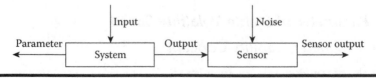

Figure 6.5 System input/output block diagram.

where the system matrix is $A \in M_n$, the output matrix is $C \in M_{m,n}$, and the parameter-state relation matrix is $P \in M_{l,n}$. The system constant vector is $b \in R^n$ $d \in R^m$, $e \in R^l$.

Definition 6.6: Given the system (6.7 through 6.10), the parameter stability domain requirement Θ, if there is a constant $K_0 \geq 0$, makes

$$r(k) \in \Theta, \ \forall k \geq K_0.$$

Then the parameter $r(k)$ can be called stable, which is also known as parameter bounded stable. Given the norm $\|\cdot\|$, if the $r^* \in R^l$ exists, for any $\varepsilon > 0$, there is the constant $K_1 \geq 0$, which makes

$$\left\| r(k) - r^* \right\| < \varepsilon, \ \forall k \geq K_1.$$

Then the parameter $r(k)$ is asymptotically stable and the stable point is r^*, namely,

$$r(k) \to r^*, \ k \to \infty.$$

Obviously, if the parameter $r(k)$ is asymptotically stable, and there is a norm $\|\cdot\|$ and constant $L > 0$, this makes

$$\left\{ x \in R^l \, \|x - r^*\| \leq L \right\} \subset \Theta.$$

Then the system is bounded stable. The main focus of this research report for the asymptotic stability of parameters unfolds. Hereinafter, the "stability" and "asymptotic stability" will not be distinguished.

Definition 6.7: Define $T \overset{\Delta}{=} \max\left\{ k \mid r(k) \notin \Theta \right\}$ as the parameter's initial volatilities; when the time is $k > T$, the system begins to enter the stable phase. In the system design, the ideal solution is (in the case of not affecting system performance) to make the initial volatility period as short as possible. In general, the system stability domain is taken as

$$\Theta \overset{\Delta}{=} \left\{ x \in R^l \mid \|x - r^*\| \leq Q \right\},$$

where $Q > 0$ is a given value.

6.2.5.2 Conditions for Determining Parameter Stability

Combining the concept of stable system parameter stability given in the first section of this chapter, before the study of various parameters' stability conditions, several lemmas should be given first.

Lemma 1 [87]: Set $A \in M_n$, and $\varepsilon > 0$ is given. Then there is at least one matrix norm $\|\cdot\|$, making

$$\rho(A) \le \|A\| \le \rho(A) + \varepsilon,$$

where $\rho(A) \overset{\Delta}{=} \max\{|\lambda| : \text{is the characteristic value of } A\}$, which is the spectral radius of matrix A.

Lemma 2 [88]: Set $\|\cdot\|_\alpha$ and $\|\cdot\|_\beta$ as any two vector norms of finite-dimensional real or complex vector space V. Then there is a constant $C_m, C_M > 0$, which makes

$$C_m \|\cdot\|_\alpha \le \|\cdot\|_\beta \le C_M \|\cdot\|_\alpha$$

true for all $x \in V$.

Theorem 6.1: For the system (6.7 through 6.10), given a constant $Q > 0$, if all the modules of the characteristic root of the system matrix A are less than 1 (namely, the spectral radius of the system matrix A is less than 1), then the system parameters are asymptotically stable.

This demonstrates that when all the characteristic root modules of the matrix A are less than 1, then the matrixes $I - A$ are reversible, the system is a stable system, and the system equilibrium point is

$$x^* = (I - A)^{-1} b.$$

Taking

$$r^* = P(I - A)^{-1} b + e,$$

due to $\rho(A) < 1$, according to Lemma 1, there must exist an induced norm $\|\cdot\|_\alpha$ meeting

$$\rho(A) \le \| A \|_\alpha \le \rho(A) + \frac{1 - \rho(A)}{2} = \frac{1 + \rho(A)}{2} < 1.$$

Then the corresponding vector norms are as follows:

$$\|r(k) - r^*\|_\alpha = \|Px(k) - Px^*\|_\alpha \le \|P\|_\alpha \|A\|_\alpha \|x(k-1) - x^*\| \le \dots \le \|P\|_\alpha \|A\|_\alpha^k \|x(0) - x^*\|_\alpha.$$

According to Lemma 2, for any norm $\|\cdot\|$ and a norm $\|\cdot\|_\alpha$, there is certainly a constant $C_M > 0$, which makes $\|\cdot\| \leq C_M \|\cdot\|_\alpha$, set up for all $x \in V$, then

$$\left\| r(k) - r^* \right\| \leq C_M \left\| r(k) - r^* \right\|_\alpha \leq C_M \|P\|_\alpha \|A\|_\alpha^k \left\| x(0) - x^* \right\|_\alpha.$$

According to Definition 6.6, the system parameter $r(k)$ is asymptotically stable.

Further, suppose matrix A's n linearly independent feature vectors $\alpha_1, \alpha_2, \ldots, \alpha_n$; the corresponding latent roots of which then are $\lambda_1, \lambda_2, \ldots, \lambda_n (|\lambda_1| \geq |\lambda_2| \geq .. \geq |\lambda_n|)$ and $(\lambda_i \neq 1)$, respectively. Because the feature vectors $\alpha_1, \alpha_2, \ldots, \alpha_n$ are linearly independent, then there is a set of constants $s_1, s_2, \ldots, s_n$ that make

$$x_0 - x^* = s_1 \alpha_1 + s_2 \alpha_2 + \ldots + s_n \alpha_n, \tag{6.11}$$

where

$$x^* = (I - A)^{-1} b$$

is the system balance point:

$$r^* = P (I - A)^{-1} b + e.$$

There is

$$r(k) - r^* = P[x(k) - x^*] = PA[x(k-1) - x^*] = \ldots = PA^k [x(0) - x^*]$$

$$= PA^k (s_1 \alpha_1 + s_2 \alpha_2 + \ldots + s_n \alpha_n)$$

$$= P(s_1 A^k \alpha_1 + s_2 A^k \alpha_2 + \ldots + s_n A^k \alpha_n)$$

$$= P(s_1 \lambda_1^k \alpha_1 + s_2 \lambda_2^k \alpha_2 + \ldots + s_n \lambda_n^k \alpha_n)$$

$$= s_1 \lambda_1^k P \alpha_1 + s_2 \lambda_2^k P \alpha_2 + \ldots + s_n \lambda_n^k P \alpha_n. \tag{6.12}$$

Then, if

$$r(k) \rightarrow r^*, \quad k \rightarrow \infty.$$

the following conditions must be met

$$s_1 \lambda_1^k P \alpha_1 + s_2 \lambda_2^k P \alpha_2 + \ldots + s_n \lambda_n^k P \alpha_n \rightarrow 0, \quad k \rightarrow \infty.$$

Thus, we can reach the following conclusions:

Theorem 6.2: For the system (6.7 through 6.10), the system matrix A's n linearly independent feature vectors are α_1, α_2, and ..., α_n, and the corresponding latent roots are $|\lambda_1| \geq ... \geq |\lambda_j| \geq 1 > |\lambda_{j+1}| \geq ... \geq |\lambda_n|$ and $(0 \leq j \leq n)$. If the following condition is met,

$$s_1 P \alpha_1 = 0, \ s_2 P \alpha_2 = 0, \ ..., \ s_j P \alpha_j = 0,$$

then the system parameter $r(k)$ is asymptotically stable. When the matrix P is an invertible matrix, the aforementioned condition becomes

$$s_1 = s_2 = 0 = ... s_j = 0.$$

Theorem 6.2 shows that for each $i = 1, 2, ..., j$ when $s_i = 0$ or α_i is the root of $P_x = 0$, the system parameter $r(k)$ is asymptotically stable. The theorem states the stability of parameters that are related to system stability as well as to the system initial values and parameter-state relations.

6.2.5.3 Parameter Steady Speed Measurement

A number of system parameter stability determining conditions have been given earlier. For the parameter stability system, how can we measure the parameter stability speed and how can we speed up the parameter's stability? To solve these problems, we need a measurement for parameter stability speed.

According to Theorem 6.2 (for the system with stable parameters) because

$$s_1 P \alpha_1 = 0, \ s_2 P \alpha_2 = 0, \ ..., \ s_j P \alpha_j = 0,$$

then

$$r(k+1) - r^* = s_{j+1} \lambda_{j+1}^{k+1} P \alpha_{j+1} + s_{j+2} \lambda_{j+2}^{k+1} P \alpha_{j+2} + ... + s_n \lambda_n^{k+1} P \alpha_n$$

$$r(k) - r^* = s_{j+1} \lambda_{j+1}^{k} P \alpha_{j+1} + s_{j+2} \lambda_{j+2}^{k} P \alpha_{j+2} + ... + s_n \lambda_n^{k} P \alpha_n.$$

Therefore,

$$\frac{\left\| r(k+1) - r^* \right\|}{\left\| r(k) - r^* \right\|} = \frac{\left\| s_{j+1} \lambda_{j+1}^{k+1} P \alpha_{j+1} + s_{j+2} \lambda_{j+2}^{k+1} P \alpha_{j+2} + ... + s_n \lambda_{j+n}^{k+1} P \alpha_{j+n} \right\|}{\left\| s_{j+1} \lambda_{j+1}^{k} P \alpha_{j+1} + s_{j+2} \lambda_{j+2}^{k} P \alpha_{j+2} + ... + s_n \lambda_{j+n}^{k} P \alpha_{j+n} \right\|}$$

$$= \lambda_{j+1} \frac{\left\| s_{j+1} P\alpha_{j+1} + s_{j+2} \left(\dfrac{\lambda_{j+2}}{\lambda_{j+1}} \right)^{k+1} P\alpha_{j+2} + \ldots + s_n \left(\dfrac{\lambda_n}{\lambda_{j+1}} \right)^{k+1} P\alpha_n \right\|}{\left\| s_{j+1} P\alpha_{j+1} + s_{j+2} \left(\dfrac{\lambda_{j+2}}{\lambda_{j+1}} \right)^{k} P\alpha_{j+2} + \ldots + s_n \left(\dfrac{\lambda_n}{\lambda_{j+1}} \right)^{k} P\alpha_n \right\|}. \qquad (6.13)$$

Obviously, the convergence rate of $\dfrac{\left\| r(k+1) - r^* \right\|}{\left\| r(k) - r^* \right\|}$ is related to the following parameter values:

$$\lambda_{j+1}, \frac{\lambda_{j+2}}{\lambda_{j+1}}, \frac{\lambda_{j+3}}{\lambda_{j+1}}, \ldots, \frac{\lambda_n}{\lambda_{j+1}}.$$

Then the following conclusions can be drawn:

1. In general, the smaller the $\left| \lambda_{j+1} \right|$ is, most of the constants k and $\dfrac{\left\| r(k+1) - r^* \right\|}{\left\| r(k) - r^* \right\|}$ will be smaller. Thus, the convergence of parameter $r(.)$ will be faster.

2. When $\dfrac{\left| \lambda_{j+2} \right|}{\left| \lambda_{j+1} \right|}, \dfrac{\left| \lambda_{j+3} \right|}{\left| \lambda_{j+1} \right|}, \ldots, \dfrac{\left| \lambda_n \right|}{\left| \lambda_{j+1} \right|}$ are smaller, the gap between $\left| \lambda_{j+1} \right|$ and $\left| \lambda_{j+2} \right|$, $\left| \lambda_{j+2} \right|$, $\left| \lambda_{j+3} \right|$, ..., $\left| \lambda_n \right|$ is greater. Then, for most of the constants, k and $\dfrac{\left\| r(k+1) - r^* \right\|}{\left\| r(k) - r^* \right\|}$ are smaller, and the convergence of parameter $r(.)$ is faster.

Obviously, $\left| \lambda_{j+1} \right|, \dfrac{\left| \lambda_{j+2} \right|}{\left| \lambda_{j+1} \right|}, \dfrac{\left| \lambda_{j+3} \right|}{\left| \lambda_{j+1} \right|}, \ldots, \dfrac{\left| \lambda_{j+n} \right|}{\left| \lambda_{j+1} \right|}, \dfrac{\left| \lambda_n \right|}{\left| \lambda_{j+1} \right|}$ can be selected as the measurement for describing the steady speed of the parameters. In engineering applications, in order to use $\left| \lambda_{j+1} \right|$ directly for describing the stable speed of the system parameters and when $\left| \lambda_{j+1} \right|$ is close to 1, taking into account that the engineering requirements for parameters stability is small parameter change, then $\left| \lambda_{j+2} \right|$ can be used as a description of the parameters' steady speed.

6.2.5.4 Definition of Instantaneous Availability Volatilities Stable Rate

Suppose that the system consists of single components, the failure time X complies with the general discrete distribution

$$p_k = P\{X = k\} \quad k = 0, 1, 2, \ldots$$

Assuming there is immediate repair after the system fails, the system is just like a new one after the repair, and the repair time Y obeys the general discrete distribution

$$q_k = P\{Y = k\} \quad k = 0, 1, 2, \dots$$

In order to distinguish different situations of the system, we define the system status as

$$\begin{cases} Z(k) = 0 & \textit{System Normal at K} \\ Z(k) = 0 & \textit{System Failure at K} \end{cases} \quad k = 0, 1, 2, \dots$$

Without a loss of generality, suppose that the components are new at the start, namely $P\{Z(0) = 0\} = 1$.

System failure rate and repair rate are

$$\lambda(k) \in (0,1), k = 0,1,2,\dots n_1 - 1, \quad \lambda(n_1) = 1$$

$$\mu(k) \in (0,1), k = 0,1,2,\dots n_2 - 1, \quad \mu(n_2) = 1$$

Then the system state transition equation is

$$P(k+1) = BP(k) \quad k = 0,1,2,\dots \tag{6.14}$$

The instantaneous availability of the system is

$$A(k) = \delta_{n_1+1, n_2+1} P(k), \tag{6.15}$$

where

$$B = \begin{bmatrix}
0 & 0 & \cdots & 0 & 0 & \mu(0) & \mu(1) & \cdots & \mu(n_2-1) & 1 \\
1-\lambda(0) & 0 & \cdots & 0 & 0 & 0 & 0 & \cdots & 0 & 0 \\
0 & 1-\lambda(1) & \cdots & 0 & 0 & 0 & 0 & \cdots & 0 & 0 \\
\cdots & \cdots & \cdots & \cdots & \cdots & \cdots & \cdots & \cdots & \cdots & \cdots \\
0 & 0 & \cdots & 1-\lambda(n_1-1) & 0 & 0 & 0 & \cdots & 0 & 0 \\
\lambda(0) & \lambda(1) & \cdots & \lambda(n_1-1) & 1 & 0 & 0 & \cdots & 0 & 0 \\
0 & 0 & \cdots & 0 & 0 & 1-\mu(0) & 0 & \cdots & 0 & 0 \\
0 & 0 & \cdots & 0 & 0 & 0 & 1-\mu(1) & \cdots & 0 & 0 \\
\cdots & \cdots & \cdots & \cdots & \cdots & \cdots & \cdots & \cdots & \cdots & \cdots \\
0 & 0 & \cdots & 0 & 0 & 0 & 0 & \cdots & 1-\mu(n_2-1) & 0
\end{bmatrix}$$

$$P(k) = (P_0(k,0), P_0(k,1), \ldots, P_0(k,n_1), P_1(k,0), P_1(k,1), \ldots, P_0(k,n_1))^T \in [0,1]^{m_1+n_2+2}$$

$$\delta_{m,n} = \left(\overbrace{1, \ldots, 1}^{m}, \overbrace{0, \ldots 0}^{n} \right).$$

To make the proper transformation, take

$$PV(k) = (P_0(k,0), P_0(k,1), \ldots, P_0(k,n_1), P_1(k,0), P_1(k,1), \ldots, P_0(k,n_1-1))^T \in [0,1]^{m_1+n_2+1}.$$

Then the following equation is true:

$$PV(k+1) = B_1 PV(k) + B_2, \tag{6.16}$$

where

$$B_1 = \begin{bmatrix}
-1 & -1 & \ldots & -1 & -1 & \mu(0)-1 & \mu(1)-1 & \ldots & \mu(n_2-2)-1 & \mu(n_2-1)-1 \\
1-\lambda(0) & 0 & \ldots & 0 & 0 & 0 & 0 & \ldots & 0 & 0 \\
0 & 1-\lambda(1) & \ldots & 0 & 0 & 0 & 0 & \ldots & 0 & 0 \\
\ldots & \ldots & \ldots & \ldots & \ldots & \ldots & \ldots & \ldots & \ldots & \ldots \\
0 & 0 & \ldots & 1-\lambda(n_1-1) & 0 & 0 & 0 & \ldots & 0 & 0 \\
\lambda(0) & \lambda(1) & \ldots & \lambda(n_1-1) & 1 & 0 & 0 & \ldots & 0 & 0 \\
0 & 0 & \ldots & 0 & 0 & 1-\mu(0) & 0 & \ldots & 0 & 0 \\
0 & 0 & \ldots & 0 & 0 & 0 & 1-\mu(1) & \ldots & 0 & 0 \\
\ldots & \ldots & \ldots & \ldots & \ldots & \ldots & \ldots & \ldots & \ldots & \ldots \\
0 & 0 & \ldots & 0 & 0 & 0 & 0 & \ldots & 1-\mu(n_2-2) & 0
\end{bmatrix}$$

$$B_2 = (1,0,\ldots 0)^T = \delta^T_{1,m_1+m_2} \in R^{m_1+n_2+1}.$$

The convergence point is

$$PV^* = (I - B_1)^{-1} B_2 \quad ((I - B_1) \text{ obviously reversible}).$$

According to the analysis in the previous section, we can set an .index describing the system state under steady speed, namely,

$$K = |\lambda_1(B_1)|. \tag{6.17}$$

Obviously, it can be proved that

$$\lim_{k \to \infty} \frac{\|PV(k+1) - PV^*\|}{\|PV(k) - PV^*\|} \leq |\lambda_1(B_1)|.$$

6.2.5.5 Instantaneous Availability Volatility Stability Simulation Analysis

Suppose that the system failure rate and repair rate are

$$\lambda(k) = 1 - q_1^{(k+1)^{\beta_1} - k^{\beta_1}}, \quad k = 0,1,2,\ldots n_1 - 1, \quad \lambda(n_1) = 1$$

$$\mu(k) = 1 - q_2^{(k+1)^{\beta_2} - k^{\beta_2}}, \quad k = 0,1,2,\ldots n_2 - 1, \quad \mu(n_2) = 1.$$

Taking $n_1 = n_2 = 595$, according to the different values of other parameters, select the following sets:

The first set:

$$q_1 = 0.998, \quad \beta_1 = 3; \quad q_2 = 0.98, \quad \beta_2 = 3$$

The maximum characteristic root module $|\lambda_1(B_1)| = 0.9046$.
The second set:

$$q_1 = 0.998, \quad \beta_1 = 2; \quad q_2 = 0.98, \quad \beta_2 = 2$$

The maximum characteristic root module $|\lambda_1(B_1)| = 0.9142$.
The third set:

$$q_1 = 0.998, \quad \beta_1 = 2; \quad q_2 = 0.98, \quad \beta_2 = 2$$

The maximum characteristic root module $|\lambda_1(B_1)| = 0.9978$.
The fourth set:

$$q_1 = 0.998, \quad \beta_1 = 1; \quad q_2 = 0.998, \quad \beta_2 = 1$$

The maximum characteristic root module $|\lambda_1(B_1)| = 0.9980$.

According to the fourth set of simulation calculations, the module of every largest characteristic root gradually increases. And from Figures 6.6 through 6.9, it can be found that the volatility time of the instantaneous availability is gradually increased. This result is completely identical with the earlier qualitative analysis conclusions. Therefore, it can be assumed that it is proper to use the greatest characteristic root model to express a fluctuant steady speed, and that this specific calculation of this model is independent from the subjective human factors, which can provide an objective equation for the volatility mechanism.

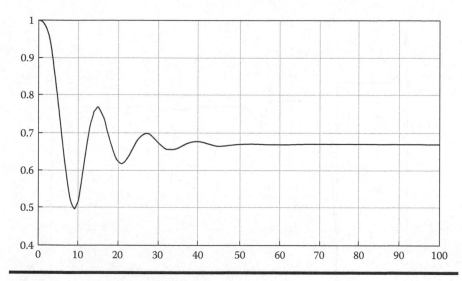

Figure 6.6   The first set of simulation examples.

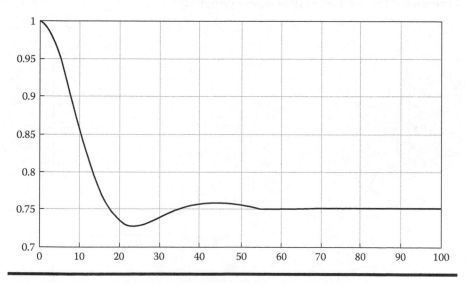

Figure 6.7   The second set of simulation examples.

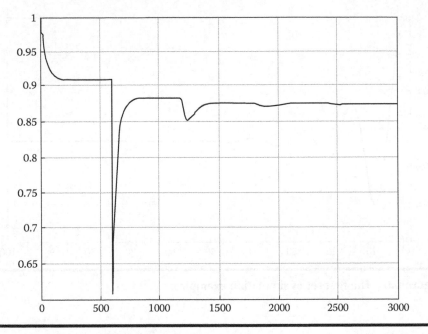

Figure 6.8 The third set of simulation examples.

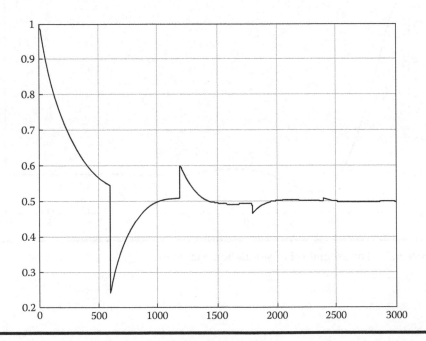

Figure 6.9 The fourth set of simulation examples.

6.3 Mechanism Analysis on Volatility of System Instantaneous Availability

The mechanism analysis can be conducted in two methods, i.e., analytical methods and numerical methods.

6.3.1 Theoretical Analysis on Volatility of System Instantaneous Availability

Five types of cases will be studied in this section.

6.3.1.1 Introduction of the Instantaneous Availability Model

To simplify the analysis, select the instantaneous availability models of the single-component two-state system, namely,

$$A(t) = P\{X(t) = 1 \mid The\ system\ is\ new\ at\ 0\}.$$

Through total probability formula:

$$A(t) = P\{X_1 > t, X(t) = 1 \mid X(0) = 1\}$$
$$+ P\{X_1 \leq t < X_1 + Y_1, X(t) = 1 \mid X(0) = 1\}$$
$$+ P\{X_1 + Y_1 \leq t, X(t) = 1 \mid X(0) = 1\};$$

obtain the updated equation

$$A(t) = 1 - F(t) + Q(t)^* A(t), \tag{6.18}$$

where

$$Q(t) = F(t)^* G(t) = \int_0^t G(t - u) \, dF(u). \tag{6.19}$$

$F(t)$ and $G(t)$ are the probability distribution functions of X and Y. The system steady-state availability is

$$A = \lim_{t \to \infty} A(t) = \frac{u}{\lambda + u}, \tag{6.20}$$

where $\dfrac{1}{\lambda} = \int_0^\infty t \, dF(t)$, $\dfrac{1}{u} = \int_0^\infty t \, dG(t)$.

6.3.1.2 Both X and Y Obey Uniform Distribution ($M_1 = M_2 = M$)

Let failure time X obeys uniform distribution for $F(t)$, and repair time Y obeys uniform distribution for $G(t)$, where the distribution functions are as follows:

$$F(t) = \begin{cases} \dfrac{t}{M} & 0 \le t \le M \\ 1 & t > M \end{cases}; \quad G(t) = \begin{cases} \dfrac{t}{M} & 0 \le t \le M \\ 1 & t > M \end{cases}, \quad (6.21)$$

where M is the parameter of uniform distribution.

By using Formula 6.21, we can draw the following conclusions:

Theorem 6.3: Let failure time X obey uniform distribution for $F(t)$, and let repair time Y obey uniform distribution for $G(t)$, where $F(t) = G(t)$. Then the analytical solution of instantaneous availability $A(t)$ when $t \in [0, 3M]$ is as follows:

$$A(t) = \begin{cases} e^{-\frac{1}{M}t}, & t \in [0, M] \\[2mm] \left(1 - e + \dfrac{t}{M}e\right)e^{-\frac{1}{M}t}, & t \in [M, 2M] \quad (6.22) \\[2mm] \left(1 - (e - 2e^2) + (e - 2e^2)\dfrac{t}{M} + \dfrac{t^2}{2M^2}e^2\right)e^{-\frac{1}{M}t}, & t \in [2M, 3M]. \end{cases}$$

Proof: Substituting the distribution functions (6.21) into Formula 6.19, the piecewise function $Q(t)$ can be obtained in the following:

$$Q(t) = \int_0^t G(t - u)dF(u) = \int_0^t G(t - u)F'(u)du$$

$$= \begin{cases} \displaystyle\int_0^t \dfrac{t - u}{M} \cdot \dfrac{1}{M}du, & t \in [0, M] \\[3mm] \displaystyle\int_0^M G(t - u)\dfrac{1}{M}du + \int_M^t G(t - u)0\,du, & t \in [M, 2M] \\[3mm] \displaystyle\int_0^M G(t - u)\dfrac{1}{M}du + \int_M^t G(t - u)0\,du, & t \in [2M, +\infty] \end{cases}$$

$$= \begin{cases} \dfrac{t^2}{2M^2}, & t \in [0,M] \\[2ex] -\dfrac{t^2}{2M^2} + \dfrac{2t}{M} - 1, & t \in [M,2M] \\[2ex] 1, & t \in [2M,+\infty] \end{cases} \qquad (6.23)$$

Then, substituting $Q(t)$ into the renewal equation (6.18), we can get:

$$A(t) = 1 - F(t) + Q(t) * A(t)$$

$$= \begin{cases} 1 - \dfrac{t}{M} + \displaystyle\int_0^t \dfrac{t-u}{M^2} A(u)\,du, & t \in [0,M] \\[3ex] \displaystyle\int_{t-M}^t \dfrac{t-z}{M^2} A(z)\,dz + \int_0^{t-M}\left(-\dfrac{t-z}{M^2} + \dfrac{2}{M}\right) A(z)\,dz, & t \in [M,2M] \ (6.24) \\[3ex] \displaystyle\int_{t-M}^t \dfrac{t-z}{M^2} A(z)\,dz + \int_{t-2M}^{t-M}\left(-\dfrac{t-z}{M^2} + \dfrac{2}{M}\right) A(z)\,dz, & t \in [2M,+\infty] \end{cases}$$

We take the derivative of Equation 6.24 to obtain the two-order delay differential equations with piecewise:

$$A''(t) = \begin{cases} \dfrac{1}{M^2} A(t), & t \in [0,M] \\[2ex] \dfrac{1}{M^2} A(t) - \dfrac{2}{M^2} A(t-M), & t \in [M,2M] \\[2ex] \dfrac{1}{M^2} A(t) - \dfrac{2}{M^2} A(t-M) + \dfrac{1}{M^2} A(t-2M), & t \in [2M,+\infty] \end{cases} \qquad (6.25)$$

For the delay differential equations with piecewise (6.25), we can get the analytical solution when $t \in [0, 3M]$ by using the initial condition $A(0) = 1$, the satisfaction of the initial integral equations, and the continuity of function $A(t)$, which are in the following:

$$A(t) = \begin{cases} e^{-\frac{1}{M}t}, & t \in [0,M] \\[2ex] \left(1 - e + \dfrac{t}{M}e\right)e^{-\frac{1}{M}t}, & t \in [M,2M] \\[2ex] \left(1 - (e - 2e^2) + (e - 2e^2)\dfrac{t}{M} + \dfrac{t^2}{2M^2}e^2\right)e^{-\frac{1}{M}t}, & t \in [2M,3M] \end{cases} \qquad (6.26)$$

According to the analytical solution of the instantaneous availability $A(t)$ in Theorem 6.3, the method to judge whether the volatility of $A(t)$ exists can be given. By using Formula 6.23, the continuity of function $Q'(t)$ is easily proven, and $1 - F(t)$ is a bounded function such that $A(t)$ is also continuous and unique. So, the method of judging the volatility of $A(t)$ is that if the value of $A(t)$ less than that of steady-state availability A exists in the finite time, it is known that the volatility of $A(t)$ exists in terms of the stability theory of availability. The steps for judging the volatility of $A(t)$ are as follows.

First, the analytical solution (6.22) is used when $t \in [0, M]$. If the value of $A(t)$ is less than that of steady-state availability A, the volatility of $A(t)$ exists. Otherwise, consider the situation of $t > M$ as follows:

Step one:
    Calculate the steady-state availability A.
    Using the Formulas 6.20 and 6.21, $A = 0.5$.

Step two:
    Judge whether the value of $A(t)$ is less than that of A when $t \in [0, M]$.

From Formula 6.22, $A(t) = e^{-\frac{t}{M}}, \quad t \in [0, M]$.

Therefore, $A(M) = e^{-1} < A = 0.5$; namely, the value of $A(t)$ less than that of A exists such that the volatility of $A(t)$ exists.

Next, the validity of this method can be illustrated by the simulations.

For Formula 6.22, make $M = 1, 5, 10, 15, 20$, respectively, and $A = 0.5$; the curves can be simulated as in Figure 6.10.

From Figure 6.10 we can see that when failure time X and repair time Y obey the same uniform distribution, the instantaneous availability $A(t)$ is less than the steady-state availability A at the time of M. This shows that the volatility of $A(t)$ indeed exists. Keep in mind that the choice of parameter M only ensures the curves are clear and complete by the comparison; in fact, the volatility of $A(t)$ exists at any parameter M.

6.3.1.3 Both X and Y Obey Uniform Distribution ($M_1 < M_2$)

Let failure time X obey uniform distribution for $F(t)$, and repair time Y obey uniform distribution for $G(t)$, where the distribution functions are as follows:

$$F(t) = \begin{cases} \dfrac{t}{M_1} & 0 \leq t \leq M_1 \\ 1 & t > M_1 \end{cases}; \quad G(t) = \begin{cases} \dfrac{t}{M_2} & 0 \leq t \leq M_2 \\ 1 & t > M_2 \end{cases} \tag{6.27}$$

where M_1 and M_2 are the parameters of uniform distribution.

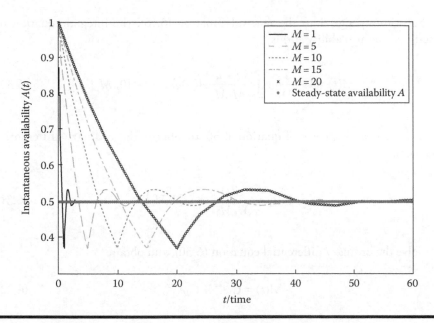

Figure 6.10 **The comparative curves of availability when *X* and *Y* obey the same uniform distribution.**

Without loss of generality, the choice of M_1 and M_2 satisfies $M_1 < M_2$, and by using Formula 6.27, we can draw the following conclusions.

Theorem 6.4: Let failure time X obey uniform distribution for $F(t)$, and repair time Y obey uniform distribution for $G(t)$, where $F(t) \neq G(t)$. Then, the analytical solution of instantaneous availability $A(t)$ when $t \in [0, M]$ is as follows:

$$A(t) = \frac{1}{2}(1-\alpha)e^{\lambda t} + \frac{1}{2}(1+\alpha)e^{-\lambda t}, \tag{6.28}$$

where $\lambda = \sqrt{\dfrac{1}{M_1 M_2}}$, $\alpha = \sqrt{\dfrac{M_2}{M_1}}$.

Proof: Substituting distribution function (6.27) into Formula 6.19, we can obtain:

$$Q(t) = \int_0^t G(t-u)\,dF(u) = \int_0^t G(t-u)F'(u)\,du$$

$$= \int_0^t \frac{t-u}{M_2} \cdot \frac{1}{M_1}\,du = \frac{t^2}{2M_1 M_2}, \quad t \in [0, M_1] \tag{6.29}$$

Substituting $Q(t)$ into the renewal equation (6.18), the integral equation of instantaneous availability $A(t)$ is

$$A(t) = 1 - \frac{t}{M_1} + \int_0^t \frac{t-u}{M_1 M_2} A(u)du, \quad t \in [0, M_1] \tag{6.30}$$

Take the derivative of Equation 6.30 to obtain the ordinary differential equation:

$$A''(t) = \frac{A(t)}{M_1 M_2}, \quad t \in [0, M_1]. \tag{6.31}$$

Solve the ordinary differential equation (6.30), and obtain

$$A(t) = C_1 e^{\lambda t} + C_2 e^{-\lambda t}, \tag{6.32}$$

where $\lambda = \sqrt{\dfrac{1}{M_1 M_2}}$; C_1 and C_2 are the undetermined coefficients.

According to the initial condition $A(0) = 1$ and the satisfaction of the initial integral equation, we can conclude

$$A(t) = \frac{1}{2}(1-\alpha)e^{\lambda t} + \frac{1}{2}(1+\alpha)e^{-\lambda t}, \tag{6.33}$$

where $\lambda = \sqrt{\dfrac{1}{M_1 M_2}}, \quad \alpha = \sqrt{\dfrac{M_2}{M_1}}$.

Using Theorem 6.4, the analytical solution of instantaneous availability $A(t)$ with $t \in [0, M_1]$ can be given when failure time X and repair time Y obey different uniform distributions. To study the existence of volatility further, we draw the following conclusion.

Theorem 6.5: When $t \in (0, 1)$, $f(t) = \dfrac{1}{2}\left(1 - \dfrac{1}{t}\right)e^t + \dfrac{1}{2}\left(1 + \dfrac{1}{t}\right)e^{-t} - \dfrac{t^2}{t^2 + 1} < 0$.

Proof: Arranging the function $f(t)$, we can get

$$f(t) = \frac{1}{2}(e^t + e^{-t}) + \frac{1}{2t}(e^{-t} - e^t) - 1 + \frac{1}{t^2 + 1}. \tag{6.34}$$

So, we can only prove $f(t) < 0$ in Formula 6.34 when $t \in (0,1)$.
First, make the Taylor expansion on e^t and e^{-t} when $t \in (0,1)$.
Namely,

$$e^t = 1 + \frac{t}{1!} + \frac{t^2}{2!} + \frac{t^3}{3!} e^\xi, \quad \xi \in (0,1);$$

$$e^{-t} = 1 - \frac{t}{1!} + \frac{t^2}{2!} - \frac{t^3}{3!} e^{-\eta}, \quad \eta \in (0,1);$$

so

$$f(t) = \frac{1}{2}\left(e^t + e^{-t}\right) + \frac{1}{2t}\left(e^{-t} - e^t\right) - 1 + \frac{1}{t^2+1}$$

$$= \frac{t^2}{2} - 1 + \frac{1}{t^2+1} + \frac{t^2}{12}(t-1)e^\xi - \frac{t^2}{12}(t+1)e^{-\eta}, \qquad (6.35)$$

$$< g(t)$$

where $g(t) = \dfrac{t^2}{2} - 1 + \dfrac{1}{t^2+1}, \in (0,1)$.

From Formula 6.35, if $g(t) < 0$, $t \in (0, 1)$, $f(t) < 0$.

We can easily conclude: when $g'(t) = t - \dfrac{2t}{(t^2+1)^2} < 0, 0 < t < \sqrt{\sqrt{2}-1}$, then $g'(t) = t - \dfrac{2t}{(t^2+1)^2} > 0, \sqrt{\sqrt{2}-1} < t < 1$.

Hence, max $g(t) < \{g(1) = 0, g(0) = 0\} = 0$, and then $g(t) = \dfrac{t^2}{2} - 1 + \dfrac{1}{t^2+1} < 0$ when $t \in (0, 1)$. Combined with Formula 6.35, we can conclude $f(t) < 0$ surely when $t \in (0, 1)$.

Using the image of function $f(t)$ as follows (z is vertical axis), shown in Figure 6.11; $f(t) < 0$, $t \in (0, 1)$ also can be seen in Figure 6.11.

Similar to the method for judging the existence of volatility in Section 6.3.1.2, the steps to judge volatility can be given according to Theorem 6.4 and Theorem 6.5.

Step one:
Calculate the steady-state availability A.
From Formulas 6.20 and 6.27, we can calculate

$$A = \frac{u}{\lambda+u} = \frac{M_1}{M_1+M_2} = \frac{\dfrac{M_1}{M_2}}{\dfrac{M_1}{M_2}+1}.$$

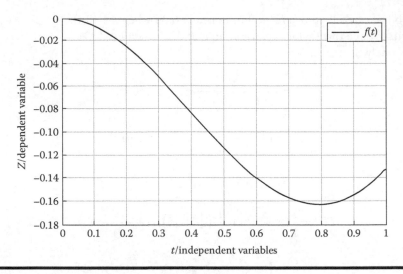

Figure 6.11 The image of functional f(t).

Step two:

Judge whether the value of $A(t)$ is less than that of A when $t \in [0, M_1]$.

From Formula 6.28, Theorem 6.5, and $\sqrt{\dfrac{M_1}{M_2}} < 1$, we can know that

$$A(M_1) - A = f\left(\sqrt{\frac{M_1}{M_2}}\right) = \frac{1}{2}\left(1 - \sqrt{\frac{M_2}{M_1}}\right)e^{\sqrt{\frac{M_1}{M_2}}} + \frac{1}{2}\left(1 + \sqrt{\frac{M_2}{M_1}}\right)e^{-\sqrt{\frac{M_1}{M_2}}} - \frac{\dfrac{M_1}{M_2}}{\dfrac{M_1}{M_2} + 1} < 0.$$

So the volatility of instantaneous availability exists.

Then, verify the validity of method by the simulation.

Using Formula 6.28, take $M_1 = 20$ and $M_2 = 40, 60, 80, 100$, respectively, and then $A = \dfrac{1}{3}, \dfrac{1}{4}, \dfrac{1}{5}, \dfrac{1}{6}$. Simulate the curves as follows: From four groups of curves in Figure 6.12, the volatility of $A(t)$ exists at one time by the continuity and marginal stability because the instantaneous availability $A(t)$ is less than the steady-state availability A when $t = M_1$. So, the theoretical analysis and simulation verify that volatility still exists under the uniform distribution for any parameter M_1, M_2.

6.3.1.4 Both X and Y Obey Exponential Distribution

In this section, we discuss the fact that the failure time and repair time of a system component both obey exponential distribution. Because the life in many electronic products obeys exponential distribution, some system life seems to be

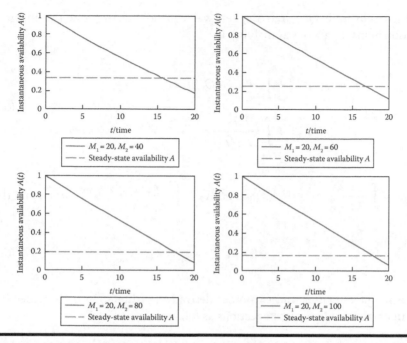

Figure 6.12 **The comparative curves of availability when *X* and *Y* obey different uniform distributions.**

used to approximate the exponential distribution. This form of distribution is most commonly used in reliability studies, so the research has a general significance.

Let failure time obey exponential distribution for $F(t)$, and repair time obey exponential distribution for $G(t)$, where the distribution functions are as follows:

$$F(t)=1-e^{-\lambda_1 t}, \quad G(t)=1-e^{-\lambda_2 t} \tag{6.36}$$

where λ_1 and λ_2 are the parameters of exponential distribution.

By using Formula 6.36, we can draw the following conclusions:

Theorem 6.6: Let failure time X obey exponential distribution for, $F(t)$ and repair time Y obey exponential distribution for $G(t)$. Then, the analytical solution of instantaneous availability $A(t)$ when $t \in (0, \infty)$ is as follows:

$$A(t) = \frac{\lambda_2}{\lambda_2 + \lambda_1} + \frac{\lambda_1}{\lambda_2 + \lambda_1} e^{-(\lambda_2 + \lambda_1)t} \tag{6.37}$$

Proof: By substituting distribution function (6.36) into (6.19), $Q(t)$ can be obtained:

$$Q(t) = \int_0^t (1-e^{-\lambda_2(t-u)})\lambda_1 e^{-\lambda_1 u}\,du = 1 - \frac{\lambda_2}{\lambda_2 - \lambda_1} e^{-\lambda_1 t} + \frac{\lambda_1}{\lambda_2 - \lambda_1} e^{-\lambda_2 t} \tag{6.38}$$

Then substitute $Q(t)$ into the renewal equation (6.18), and the integrated equation of instantaneous availability is

$$A(t) = e^{-\lambda_1 t} + \int_0^t A(t-u)Q'(u)\,du$$

$$= e^{-\lambda_1 t} + \int_0^t \frac{\lambda_1 \lambda_2}{\lambda_2 - \lambda_1}(e^{-\lambda_1(t-u)} - e^{-\lambda_2(t-u)})A(u)\,du$$

(6.39)

Let $X_1 = \int_0^t \frac{\lambda_1 \lambda_2}{\lambda_2 - \lambda_1}e^{-\lambda_1(t-u)}A(u)\,du$ and $X_2 = \int_0^t \frac{\lambda_1 \lambda_2}{\lambda_2 - \lambda_1}e^{-\lambda_2(t-u)}A(u)\,du$, and then

$$A(t) = e^{-\lambda_1 t} + X_1 - X_2. \tag{6.40}$$

Using the one-order and two-order derivatives of Equation 6.40, respectively, we can obtain two integrated equations as follows:

$$A'(t) = -\lambda_1 e^{-\lambda_1 t} - \lambda_1 X_1 + \lambda_2 X_2$$

$$A''(t) = \lambda_1^2 e^{-\lambda_1 t} + \frac{\lambda_1 \lambda_2}{\lambda_2 - \lambda_1}\left(-\lambda_1 A(t) + \lambda_2 A(t)\right) + \lambda_1^2 X_1 - \lambda_2^2 X_2.$$

(6.41)

Eliminating X_1 and X_2 in Formula 6.41, the ordinary differential equation can be deduced:

$$A''(t) + (\lambda_1 + \lambda_2)\,A'(t) = 0. \tag{6.42}$$

Solve the ordinary differential equation (6.42) with the initial condition $A(0) = 1$ and the satisfaction of the initial integrated equation, inferring

$$A(t) = \frac{\lambda_2}{\lambda_2 + \lambda_1} + \frac{\lambda_1}{\lambda_2 + \lambda_1}e^{-(\lambda_2 + \lambda_1)t},\quad t \in (0,\infty) \tag{6.43}$$

Judging from Theorem 6.6, when failure time X and repair time Y both obey exponential distribution, the analytical solution (6.37) of $A(t)$ is the exponential function so that there is no volatility of $A(t)$.

For example: As is seen in Figure 6.13, there is no volatility of $A(t)$ when X and Y both obey exponential distributions.

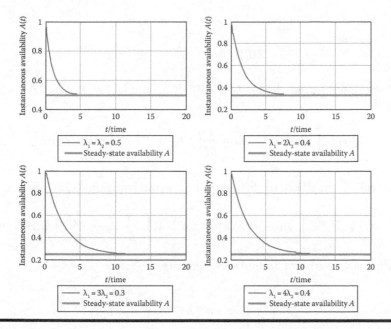

Figure 6.13 The comparative curves of availability when *X* and *Y* obey exponential distributions.

6.3.1.5 Both X and Y Obey Exponential and Uniform Distribution

This section outlines the fact that failure time obeys exponential distribution, and repair time obeys uniform distribution in the system of a single component. Because the failure and repair of a system component are independent and subject to different distributions, we selected the combination of exponential distribution and uniform distribution to study the volatility.

Let failure time obey exponential distribution for $F(t)$, and repair time obey uniform distribution for $G(t)$, where the distribution functions are as follows:

$$F(t) = 1 - e^{-\lambda t}, \ G(t) = \frac{t}{M} \tag{6.44}$$

where λ and M are the parameters of different distribution noted earlier.

By using Formula 6.44, we can draw the following conclusions:

Theorem 6.7: Let failure time X obey exponential distribution for $F(t)$ and repair time Y obey uniform distribution for $G(t)$. Then, the analytical solution of instantaneous availability $A(t)$ when $t \in [0, M]$ is as follows:

$$A(t) = \frac{(\beta_1 e^{\beta_1 t} - \beta_2 e^{\beta_2 t})}{\sqrt{\lambda^2 + \dfrac{4\lambda}{M}}}, \ t \in [0, M] \tag{6.45}$$

where $\beta_1 = \dfrac{-\lambda + \sqrt{\lambda^2 + \dfrac{4\lambda}{M}}}{2} > 0, \beta_2 = \dfrac{-\lambda - \sqrt{\lambda^2 + \dfrac{4\lambda}{M}}}{2} < 0$.

Proof: Substituting distribution function (6.44) into (6.19), $Q(t)$ can be obtained:

$$Q(t) = \int_0^t \frac{t-u}{M} \lambda e^{-\lambda u}\, du = \frac{1}{M}\left(t + \frac{1}{\lambda}e^{-\lambda t} - \frac{1}{\lambda}\right), \quad t \in [0, M]. \tag{6.46}$$

And substituting Formula 6.46 into the renewal equation (6.18), such that

$$A(t) = R(t) + \int_0^t A(t-u)Q'(u)\, du = e^{-\lambda t} + \int_0^t \frac{1}{M}(1 - e^{-\lambda(t-u)})A(u)\, du \tag{6.47}$$

Take the derivative of Formula 6.47, getting

$$A'(t) = -\lambda e^{-\lambda t} + \int_0^t \frac{\lambda}{M} e^{-\lambda(t-u)} A(u)du = -\lambda e^{-\lambda t} + X_1$$

$$A''(t) = \lambda^2 e^{-\lambda t} + \frac{\lambda}{M}A(t) - \int_0^t \frac{\lambda^2}{M}e^{-\lambda(t-u)}A(u)du = \lambda^2 e^{-\lambda t} - \lambda X_1 + \frac{\lambda}{M}A(t), \tag{6.48}$$

where $X_1 = \displaystyle\int_0^t \frac{\lambda}{M}e^{-\lambda(t-u)}A(u)\, du$.

Eliminating X_1 from Formula 6.48, we can obtain the ordinary differential equation:

$$A''(t) + \lambda A'(t) - \frac{\lambda}{M}A(t) = 0, \ t \in [0, M]. \tag{6.49}$$

Solving Equation 6.49 with the initial condition $A(0) = 1$ and the satisfaction of the initial integrated equation, we conclude that

$$A(t) = \frac{\left(\beta_1 e^{\beta_1 t} - \beta_2 e^{\beta_2 t}\right)}{\sqrt{\lambda^2 + \dfrac{4\lambda}{M}}}, \quad t \in [0, M], \tag{6.50}$$

where $\beta_1 = \dfrac{-\lambda + \sqrt{\lambda^2 + \dfrac{4\lambda}{M}}}{2} > 0, \beta_2 = \dfrac{-\lambda - \sqrt{\lambda^2 + \dfrac{4\lambda}{M}}}{2} < 0$.

Similar to the method for judging the existence of volatility in Section 6.3.1.2, the steps to judge the volatility can be given according to Theorem 6.7.

Step one:

Calculate the steady-state availability A.

From the Formulas 6.20 and 6.44, we can calculate

$$A = \frac{\dfrac{2}{M}}{\lambda + \dfrac{2}{M}}.$$

Step two:

Judge whether the extreme point of $A(t)$ exists when $t \in [0,M]$. Because

$$A(t) = \frac{(\beta_1 e^{\beta_1 t} - \beta_2 e^{\beta_2 t})}{\sqrt{\lambda^2 + \dfrac{4\lambda}{M}}}, \quad t \in [0, M],$$

solve the extreme point of $A(t)$.

Namely, $A'(t) = \beta_1^2 e^{\beta_1 t} - \beta_2^2 e^{\beta_2 t} = 0$.

Get the stagnation point:

$$t_0 = \frac{2}{\beta_1 - \beta_2} \ln\left|\frac{\beta_2}{\beta_1}\right| = \frac{2}{\sqrt{\lambda^2 + \dfrac{4\lambda}{M}}} \ln\left| \frac{M\left(\lambda + \sqrt{\lambda^2 + \dfrac{4\lambda}{M}}\right) + 2}{2} \right| \tag{6.51}$$

We easily can prove $A'(t) < 0$, $t < t_0$; $A'(t) > 0$ $t > t_0$, so t_0 is the extreme point of $A(t)$. If $t_0 < M$, the volatility of $A(t)$ exists. Therefore, we introduce Theorem 6.8 to prove $t_0 < M$.

Theorem 6.8: Function $h(t) = \ln\left(\dfrac{t + \sqrt{t^2 + 4t}}{2} + 1\right) - \dfrac{\sqrt{t^2 + 4t}}{2} < 0, (t > 0)$

Proof: Take the derivative of $h(t)$, then

$$h'(t) = \frac{1 + \dfrac{t+2}{\sqrt{t^2+4t}}}{t + 2 + \sqrt{t^2+4t}} - \frac{t+2}{2\sqrt{t^2+4t}} = \frac{-t}{2\sqrt{t^2+4t}} < 0, \quad (t > 0).$$

So $h(t) < h(0) = 0$, $(t>0)$.

From the expression (6.51), we can deduce

$$t_0 - M < \frac{2}{\sqrt{\lambda^2 + \frac{4\lambda}{M}}} \ln \left| \frac{M\left(\lambda + \sqrt{\lambda^2 + \frac{4\lambda}{M}}\right) + 2}{2} - M \right|$$

$$= \frac{2}{\sqrt{\lambda^2 + \frac{4\lambda}{M}}} \left(\ln\left(\frac{M\lambda + \sqrt{(M\lambda)^2 + 4M\lambda}}{2} + 1 \right) - \frac{\sqrt{(M\lambda)^2 + 4M\lambda}}{2} \right).$$

$$= \frac{2}{\sqrt{\lambda^2 + \frac{4\lambda}{M}}} h(M\lambda) < 0, \ M\lambda > 0$$

So $t_0 < M$, and then the volatility of $A(t)$ exists at one time when $t \in [0, M]$.

As is seen in Figure 6.14, volatility of $A(t)$ exists when X obeys exponential distribution and Y obeys uniform distribution, which is in accordance with the theoretical certification.

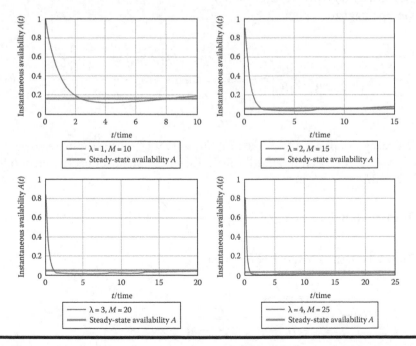

Figure 6.14 **The comparative curves of availability when X and Y obey exponential and uniform distributions.**

Table 6.1   The Conclusions of Volatility

F: uniform distribution; G: uniform distribution;	Because uniform distribution is the piecewise function, the $A(t)$ obtained is also the piecewise function. By calculating the first section of the analytic expression, the conclusion that there exists a point satisfying $A(t) < A$ is easily obtained, and it is not related to the parameters selected. Through the property of the steady-state availability, it can be introduced that $A(t)$ at least has one volatility (when t tends to infinity).
F: exponential distribution; G: exponential distribution;	From the renewal equation, we can obtain the global analytical solution. That shows that $A(t)$ has no volatility because the solution is an exponential function.
F: exponential distribution; G: uniform distribution;	$A(t)$ is also the piecewise function, and it can be proven that there exists an extreme point t_0 in the first section of the analytic expression. Therefore, $A(t)$ has at least one volatility.

6.3.1.6 Summary of Theoretical Analysis of Volatility

At last, the five analytical methods are compared, as shown in Table 6.1.

6.3.2 Numerical Methods on the Volatility of Instantaneous Availability

6.3.2.1 Numerical Algorithm

Using the analytical analysis in the previous section, the conditions for the volatility of instantaneous availability under several special distributions are proven (i.e., when both the failure rate and repair rate are constants, there is no volatility). But a general conclusion cannot be determined. Due to the restrictions of the analytical research tool, further analytical analysis is difficult to carry out in depth. Therefore, the numerical analysis method will be adopted for research. To facilitate this research, the single-component two-state system instantaneous availability model will still be adopted in Section 6.3.1.1.

The general analytical solution of instantaneous availability $A(t)$ cannot be given exactly from Formula 6.18. So, we must use two composite trapezoidal formulas to Formula 6.18 with two convolutions, and then turn Formula 6.18 into the discrete equations so that the discrete values of $A(t)$ can be obtained.

From Formula 6.18, we know that

$$A(t) = 1 - F(t) + \int_0^t Q'(t-u)A(u)\,du, \tag{6.52}$$

and, from Formula 6.19, we deduce that

$$Q'(t) = \int_0^t G'(t-u)F'(u)\,du = \int_0^1 G'(t-tz)F'(tz)\,t\,dz. \quad (6.53)$$

In the following, the steps for solving system instantaneous availability are given by the numerical method, which uses the composite trapezoidal formula for (6.52) and (6.53) orderly.

Step one:
In the interval $[a, b]$, obtain $t_k = a + kh$, $(k = 0, 1, ..., 150, h = 0.1, a = 0)$.

Step two:
Apply the composite trapezoidal formula to each $Q'(t_k)$, when $|T_{n+1}^k - T_n^k| < 0.001$,

$$Q'(t_k) \approx \tilde{Q}'(t_k) = T_{n+1}^k, \quad (6.54)$$

where

$$T_n = \frac{h}{2}\left[f(a) + 2\sum_{k=1}^{n-1} f(x_k) + f(b) \right].$$

Step three:
For Formula 6.52, fix the value of t and divide the interval $[0, t]$ equally into n parts, that is to say, assuming $h = \dfrac{t}{n}$, $a = 0$, $u_k = a + kh = kh, (k = 0, 1, \ldots, n)$.

Step four:
Use the composite trapezoidal formula:

$$\int_0^t Q'(t-u)A(u)\,du \approx \frac{h}{2}\left[Q'(t-u_0)A(u_0) + 2\sum_{k=1}^{n-1} Q'(t-u_k)A(u_k) + Q'(t-u_n)A(u_n) \right].$$
$$(6.55)$$

Substitute Formula 6.55 into Formula 6.52:

$$A(t) \approx \tilde{A}(t) = 1 - F(t) + \frac{h}{2}\left[Q'(t-u_0)A(u_0) + 2\sum_{k=1}^{n-1} Q'(t-u_k)A(u_k) + Q'(t-u_n)A(u_n) \right].$$
$$(6.56)$$

Step five:
Obtain $n = 150$, $h = 0.1$ and suppose $t = t_k = u_k$ $(k = 0, 1, ..., n)$.

This leads to the following equations according to Formula 6.56:

$$
\begin{cases}
\tilde{A}(u_0) = 1 \\
\tilde{A}(u_1) = 1 - F(u_1) + \dfrac{h}{2}\left(\tilde{Q}'(u_1) A(u_0) + \tilde{Q}'(u_0) A(u_1) \right) \\
\qquad\qquad\vdots \\
\tilde{A}(u_n) = 1 - F(u_n) + h\left(\dfrac{1}{2}\tilde{Q}'(u_n) A(u_0) + \tilde{Q}'(u_{n-1}) A(u_1) + \ldots \right. \\
\qquad\left. + \tilde{Q}'(u_1) A(u_{n-1}) + \dfrac{1}{2}\tilde{Q}'(u_0) A(u_n) \right)
\end{cases}
\tag{6.57}
$$

Among them, obtaining $\tilde{Q}'(u_k)(k = 0, 1, \ldots, n)$ can be calculated by using Formula 6.54, so the equations can be transformed into a matrix computation. That is to say:

$$
\tilde{Q} * \tilde{A} = I - F
\tag{6.58}
$$

Therefore,
$$
\tilde{A} = \begin{pmatrix} \tilde{A}(u_0) \\ \tilde{A}(u_1) \\ \vdots \\ \tilde{A}(u_n) \end{pmatrix}, \quad I - F = \begin{pmatrix} 1 - F(u_0) \\ 1 - F(u_1) \\ \vdots \\ 1 - F(u_n) \end{pmatrix},
$$

$$
\tilde{Q} = \begin{pmatrix}
1 & 0 & \cdots & 0 \\
-\dfrac{h}{2}\tilde{Q}'(u_1) & 1 & \cdots & 0 \\
\vdots & \vdots & \ddots & \vdots \\
-\dfrac{h}{2}\tilde{Q}'(u_n) & -h\tilde{Q}'(u_{n-1}) & \cdots & 1
\end{pmatrix}
$$

Step six:

After settling (6.58), obtain $\tilde{A} = \tilde{Q}^{-1}(I - F)$ from Formula 6.32, and then solve $\tilde{A}(u_k), (k = 0, 1, \ldots, n)$.

Due to the use of the two numerical integrations, $A(t)$ produces two errors that appear in computing (6.54) and in computing (6.56). The error range is discussed in the following.

Using Formula 6.54, the error of $\tilde{Q}'(t_k)$ is 0.001 and the error of composite trapezoidal formula is the order of $h^2 = 0.01$.

So that

$$\tilde{A}(t) = 1 - F(t) + \frac{h}{2}\left[\tilde{Q}'(t - u_0)\tilde{A}(u_0) + 2\sum_{k=1}^{n-1}\tilde{Q}'(t - u_k)\tilde{A}(u_k) + \tilde{Q}'(t - u_n)\tilde{A}(u_n)\right].$$

(6.59)

$$\tilde{A}(t) < \overline{A}(t) = 1 - F(t) + \frac{h}{2}\left[\begin{array}{l}\left(\tilde{Q}'(t - u_0) + 0.001\right)\tilde{A}(u_0) + 2\sum_{k=1}^{n-1}\left(\tilde{Q}'(t - u_k)\right. \\ + 0.001\right)\tilde{A}(u_k) + \left(\tilde{Q}'(t - u_n) + 0.001\right)\tilde{A}(u_n)\end{array}\right] + O(0.01)$$

(6.60)

$$\tilde{A}(t) > \overline{A}(t) = 1 - F(t) + \frac{h}{2}\left[\begin{array}{l}\left(\tilde{Q}'(t - u_0) - 0.001\right)\tilde{A}(u_0) + 2\sum_{k=1}^{n-1}\left(\tilde{Q}'(t - u_k)\right. \\ - 0.001\right)\tilde{A}(u_k) + \left(\tilde{Q}'(t - u_n) - 0.001\right)\tilde{A}(u_n)\end{array}\right] - O(0.01).$$

(6.61)

Among them, $\hat{A}(t)$ is the upper value of error and $\overline{A}(t)$ is the lower value of error.

From the error theory of the composite trapezoidal formula, we conclude

$$\tilde{A}(t) - O\left(h^2\right) \le A(t) \le \tilde{A}(t) + O\left(h^2\right). \tag{6.62}$$

Therefore,

$$|A(t) = \tilde{A}(t)| \le |\hat{A}(t) - \overline{A}(t)| = \left[\frac{h}{2}(0.002)\tilde{A}(u_0) + 2\sum_{k=1}^{n-1}(0.002)\tilde{A}(u_k) + (0.002)\tilde{A}(u_n)\right]$$

$$+ O(0.01) \le \int_0^t 0.002\,\tilde{A}(u)\,du + O(0.01) \le 0.002t + O(0.01) \le 0.03 + O(0.01), t \in [0,15].$$

As long as variable t is obtained correctly, the error of instantaneous availability is controllable. Though the error increases with the increase of variable t, the error of the actual results is much smaller than the theoretical results. From that, only take $t_{max} = 15$. Because the volatility of system instantaneous availability occurs in the early stage, obtaining $t_{max} = 15$ can also determine the volatility.

6.3.2.2 Numerical Calculation Solutions and Results

According to the numerical algorithm in the previous section, the simulation can produce the curve of instantaneous availability $A(t)$ under the probability distribution given. So $A(t)$ is more likely to have volatility in situations that can be analyzed by the experiment design.

From Formula 6.18, instantaneous availability $A(t)$ is decided by the probability distribution of failure time X and repair time Y. Then there exists the relationship of the one-to-one correspondence between the failure rate and the repair rate and their corresponding probability distribution. So the system instantaneous availability can be equivalently considered as only determined by the failure rate and repair rate; that is to say, the volatility problem of system instantaneous availability is equivalent to that of the functions of the failure rate and repair rate.

When $\lambda(t) = \lambda$ and $\mu(t) = \mu$, $F(t) = 1-e^{-\lambda t}$ and $G(t) = 1-e^{-\mu t}$. From Formula 6.18, instantaneous availability can be calculated, and the solution is $A(t) = \dfrac{\mu}{\mu+\lambda} + \dfrac{\lambda}{\mu+\lambda} e^{-(\mu+\lambda)t}$. At that moment, $A(t)$ has no volatility. When $\lambda(t)$ and $\mu(t)$ are not constants, however, instantaneous availability may have volatility.

Because of the function $\lambda(t)$, $\mu(t) \geq 0$ the two functions may be less than zero when $\lambda(t)$ and $\mu(t)$ are decreasing functions or wave functions. In Table 6.2, simulation experiments are outlined where $\lambda(t)$ and $\mu(t)$ are increasing functions in order to show the volatility of $A(t)$.

The results of the simulations obtained in Table 6.2 using Schemes 1 through 8 are shown in Figures 6.15 through 6.18.

The simulation results of Schemes 1 through 8 show that instantaneous availability has volatility when failure rate and repair rate are increasing functions, and that there is more significant volatility when the order of function is higher. Considering the analytical results, it can be determined that as long as the failure

Table 6.2 Experiment Design

Scheme	Failure Rate Variation	Repair Rate Variation
1	Constant	$at + b$ (Ascending)
2		$at^{\alpha} + b$ (Ascending)
3	$at + b$ (Ascending)	Constant
4	$at^{\alpha} + b$ (Ascending)	
5	$at + b$ (Ascending)	$at + b$ (Ascending)
6		$at^{\alpha} + b$ (Ascending)
7	$at^{\alpha} + b$ (Ascending)	$at + b$ (Ascending)
8		$at^{\alpha} + b$ (Ascending)

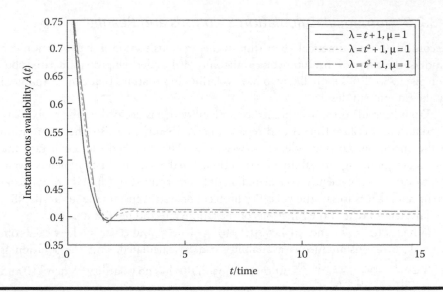

Figure 6.15 **The comparative curves of Schemes 1 through 2.**

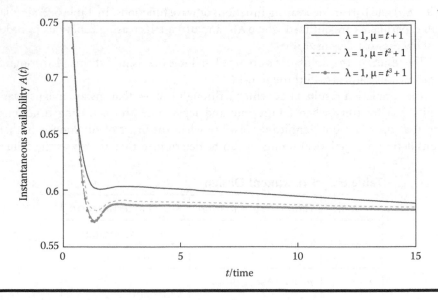

Figure 6.16 **The comparative curves of Schemes 3 through 4.**

rate or repair rate is time variant, the system instantaneous availability may have volatility. To verify this conclusion, further numerical calculation solutions should be designed.

Use $a = 1$, 0.1, 0.001, respectively, in the design to verify whether the volatility becomes weaker when $a \to 0$. Schemes 1 through 4 are the combination of

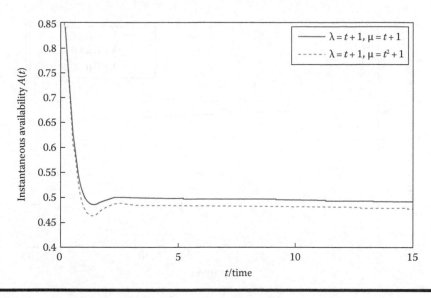

Figure 6.17   The comparative curves of Schemes 5 through 6.

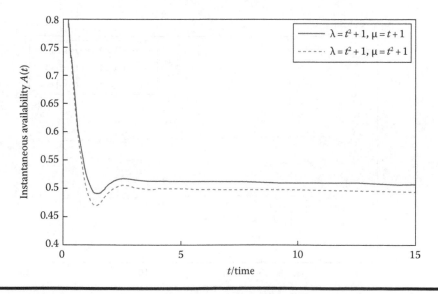

Figure 6.18   The comparative curves of Schemes 7 through 8.

constant and function, which belong to the single factor. Schemes 5 through 8 are the combinations among functions, which belong to the double factors. So Schemes 5 through 8 use the comprehensive experimental design to determine the influence of the results of the numerical calculations with the change of coefficients. The results of the simulation are shown in Figures 6.19 through 6.30.

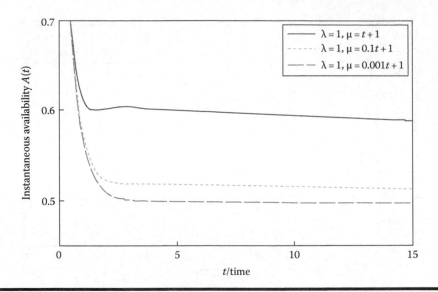

Figure 6.19   The comparative curves of Scheme 1.

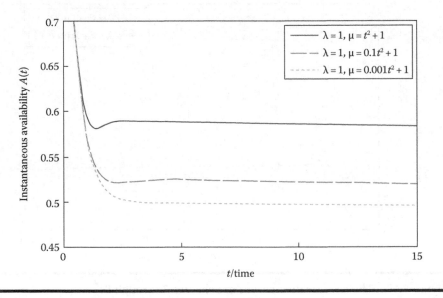

Figure 6.20   The comparative curves of Scheme 2.

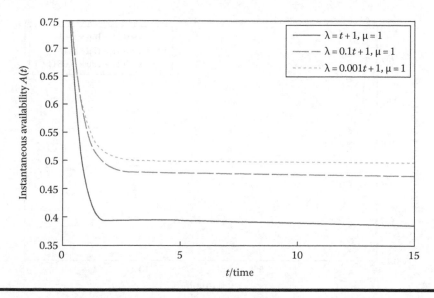

Figure 6.21   The comparative curves of Scheme 3.

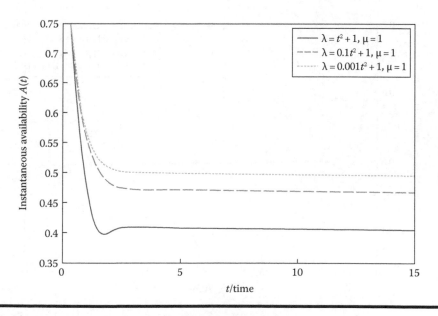

Figure 6.22   The comparative curves of Scheme 4.

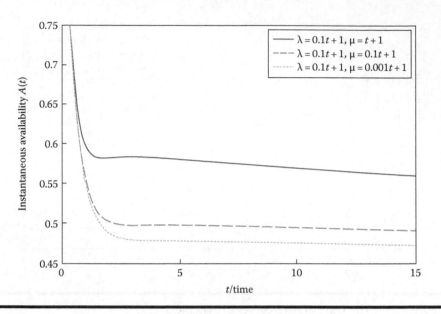

Figure 6.23   The comparative curves of Scheme 5.1.

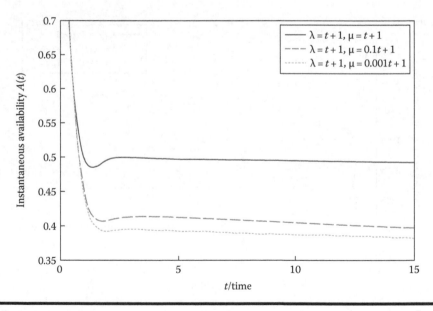

Figure 6.24   The comparative curves of Scheme 5.2.

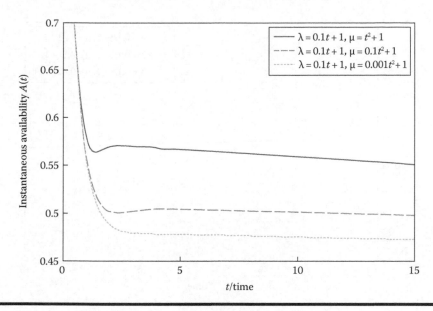

Figure 6.25   The comparative curves of Scheme 6.1.

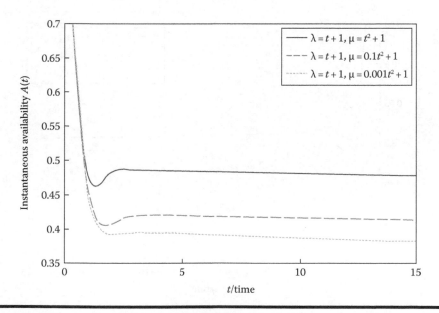

Figure 6.26   The comparative curves of Scheme 6.2.

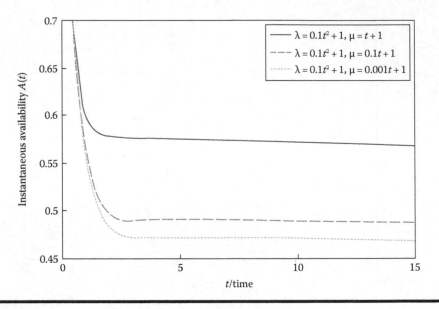

Figure 6.27   The comparative curves of Scheme 7.1.

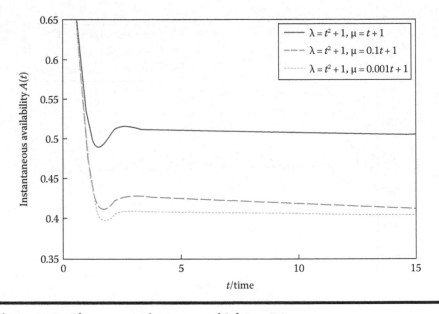

Figure 6.28   The comparative curves of Scheme 7.2.

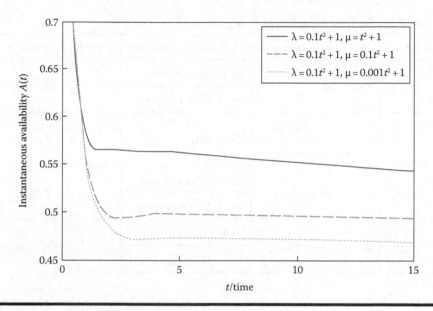

Figure 6.29   The comparative curves of Scheme 8.1.

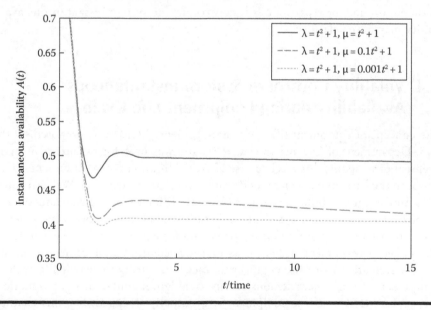

Figure 6.30   The comparative curves of Scheme 8.2.

Table 6.3 Further Experiment Design

Scheme	Failure Rate Variation	Repair Rate Variation	Parameter Variation
1	Constant	$at + 1$ (Ascending)	$a \rightarrow 0$
2		$at^2 + 1$ (Ascending)	
3	$at + 1$ (Ascending)	Constant	$a \rightarrow 0$
4	$at^2 + 1$ (Ascending)		
5	$at + 1$ (Ascending)	$at + 1$ (Ascending)	$a \rightarrow 0$
6		$at^2 + 1$ (Ascending)	
7	$at^2 + 1$ (Ascending)	$at + 1$ (Ascending)	$a \rightarrow 0$
8		$at^2 + 1$ (Ascending)	

The calculation results of Schemes 1 through 8 in Table 6.3 show that with $a \rightarrow 0$, the volatility of system instantaneous availability range will gradually decrease until it disappears. Therefore, this verifies the conclusion that as long as the failure rate or repair rate is time variant, the system instantaneous availability may have volatility.

6.4 Volatility Control of System Instantaneous Availability during Equipment Life Cycle

The concept of equipment life cycle management (LCM) was proposed by the U.S. Department of Defense in the 1970s and was used for procurement management of weapons; this was later used by the UK and France. At present, LCM has been the basic management mode for weapon procurement in Western countries. In December 2000, the China's Central Military Commission issued regulations for equipment of the Chinese People's Liberation Army, which clearly stipulates that system and LCM is carried out for equipment of the entire army. The LCM of equipment is defined as making scientific management decisions using systematic theory and engineering methods during the whole life cycle of equipment—from project demonstration to design, manufacturing, production, use, and onward until the equipment is eventually scrapped. For this chapter, following the procedures of the Chinese People's Liberation Army, the whole life cycle of equipment is divided into three phases for research purposes: the demonstration phase, the development phase, and the operation (including supporting) phase.

System availability refers to the comprehensive measurements of reliability, maintainability, and supportability of the equipment system. System availability is determined by system failure rate, repair rate, logistic delay rate, and other mathematically expressed indexes. In addition to being self-restrained, such indexes are still affected by design, application, or other factors. In addition, different phases are affected by different factors. For example, the demonstration of system failure rate, repair rate, and logistic delay rate is affected by design factors; failure rate and repair rate are also affected by design factors in the development phase. However, the design factors in the two phases are different. Unlike the previous two phases, the failure rate and repair rate are affected by operation factors in the operation phase. Furthermore, the economic cost constraint of each phase may also be different.

6.4.1 State Equation and Constraint Equation

6.4.1.1 Demonstration Phase

State equations are as follows:

$$P(k+1) = B(\lambda, \mu, \rho)P(k)$$

$$P(0) = \left(\delta_{1,m_1+n_2+n_3+2}\right)^T$$

$$A(k) = \delta_{m_1+1,n_2+n_3+2}P(k) \qquad k = 0,1,2,\ldots$$

Suppose both the inherent availability index A_{i0} and cost constraint C_1 are given; system inherent availability and cost functions have to satisfy the following formulas, respectively:

$$A_i(\lambda, \mu, \rho) \geq A_{i0}.$$

$$C_1(\lambda, \mu, \rho) \leq C_1.$$

In addition, failure rate, repair rate, and logistic delay rate have the constraint condition $(\lambda, \mu, \rho) \in \Theta_1$, and Θ_1 is affected by demonstration factors. Suppose such demonstration factors satisfy $x_1 \in \Omega_1'$. Then, the constraint condition during this phase is as follows:

$$\Omega_1 = \left\{ (\lambda, \mu, \rho) \in \Theta(x_1) \mid A_i(\lambda, \mu, \rho) \geq A_{i0}, x_1 \in \Omega_1' \right\} \cap \left\{ x \mid C_1(x) \leq C_1 \right\}.$$

6.4.1.2 Development Phase

Status equations are as follows:

$$P(k + 1) = B(\lambda, \mu)P(k)$$

$$P(0) = \left(\delta_{1, m_1 + n_2 + 1}\right)^T$$

$$A(k) = \delta_{m+1, n_2+1}P(k) \qquad k = 0, 1, 2, \ldots$$

Suppose mean time between failures $MTBF_0$, mean time to repair $MTTR_0$, and cost constraint C_2 are all given; $MTBF$, $MTTR$, and cost functions have to satisfy following formulas, respectively:

$$MTBF(\lambda) \geq MTBF_0$$

$$MTTR_0(\mu) \leq MTTR_0$$

$$C_2(\lambda, \mu) \leq C_2.$$

In addition, failure rate and repair rate have the constraint condition $(\lambda, \mu) \in \Theta_2$, and Θ_2 is affected by design factors. Suppose the vector of such design factors is $x_2 \in \Omega_2'$. Then, the constraint condition during this phase is as follows:

$$\Omega_2 = \left\{ (\lambda, \mu) \in \Theta_2(x_2) \mid MTBF(\lambda) \geq MTBF_0, MTTR(\mu) \leq MTTR_0, x_2 \in \Omega_2' \right\}$$

$$\cap \left\{ x \mid C_2(x) \leq C_2 \right\}.$$

6.4.1.3 Operation Phase

In general, state equations are as follows:

$$P(k + 1) = B(\lambda, \mu, \rho)P(k)$$

$$P(0) = \left(\delta_{1, m_1 + n_2 + n_3 + 2}\right)^T$$

$$A(k) = \delta_{m+1, n_2 + n_3 + 2}P(k) \qquad k = 0, 1, 2, \ldots$$

Suppose the cost constraint C_3 is given, and cost function has to satisfy the following formula:

$$C_3(\lambda, \mu, \rho) \leq C_3.$$

In addition, failure rate, repair rate, and logistic delay rate have the constraint condition $(\lambda, \mu, \rho) \in \Theta_3$, and Θ_3 is affected by operation factors. Suppose such operation factors satisfy $x_3 \in \Omega_3'$. Then, the constraint condition during this phase is as follows:

$$\Omega_3 = \left\{ (\lambda, \mu, \rho) \in \Theta_3(x_3) \mid x_3 \in \Omega_3' \right\} \cap \left\{ x \mid C_3(x) \le C_3 \right\}.$$

When preventive maintenance is considered, state equations are as follows:

$$P(k+1) = B(\lambda, \mu_1, \mu_2, n_1) \, P(k)$$

$$P(0) = \left(\delta_{1, m_1 + n_2 + n_3 + 2} \right)^T$$

$$A(k) = \delta_{m_1 + 1, n_2 + n_3 + 2} P(k) \qquad k = 0, 1, 2, \ldots$$

In the formula, n_1 is the preventive maintenance cycle of the system. In addition, failure rate, corrective maintenance rate, and preventive maintenance rate have the constraint condition $(\lambda, \mu_1, \mu_2, n_1) \in \Theta_3$, and Θ_3 is affected by operation factors. Suppose the vector of such operation factors is $x_3 \in \Omega_3'$. Then, the constraint condition during this phase is as follows:

$$\Omega_3 = \left\{ (\lambda, \mu_1, \mu_2, n_1) \in \Theta_3(x_3) \mid x_3 \in \Omega_3' \right\} \cap \left\{ x \mid C_3(x) \le C_3 \right\}.$$

6.4.2 Control Problem

Summarizing models of all of the aforementioned phases into one uniform model:

$$\text{Control: } I = I\,(A(.))$$

$$s.t. \qquad\qquad P(k) = B(u)P(k-1) \tag{6.63}$$

$$s.t. \qquad\qquad P(0) = \delta_2 \tag{6.64}$$

$$A(k) = \delta_1 \, P(k) \qquad k = 0, 1, 2, \ldots \tag{6.65}$$

In this case, B, δ_1, δ_2, and design variable u depend on the type of repairable system, which has been discussed in the system instantaneous availability model of each phase earlier in this chapter.

Definition 6.8 (matchability): The system (6.3 through 6.6) is given; if there is $u \in \Omega$, the availability indicative metric function $J(T) = 0$. In this case, the system is matchable at the time T; if the system is matchable at the time T for arbitrary $T \ge 0$, the system is totally matchable.

Let

$$T^* = \inf_{T \in Z^+} \left\{ T \mid \text{\textit{System is matchable at the time of}} \ T - \right\}.$$

Then, the matchable interval is defined as $[T^*, +\infty)$.

People always expect that the instantaneous availability of all new equipment systems can reach a specified requirement as soon as possible (i.e., its adaptation time is short as far as possible or volatility is slight as far as possible). This problem can be converted into an optimum control problem with a constraint.

$$\text{OPT:} \ I = I \left(A(.) \right) \tag{6.66}$$

s.t., Formulas 6.63 through 6.65 are established.

Obviously, here, the optimum control problem is different from a typical optimization problem. The aim of a typical optimization problem is just to obtain a group of static parameter points. However, we want to obtain the functional extreme value of time and the optimal solution would be the optimal function within the full time period not just a static parameter point. Therefore, this problem is related to optimum control and more difficult than a typical optimization problem.

6.5 Summary

In this chapter, research on volatility of system instantaneous availability at the initial operation stage of new equipment is carried out to propose a group of characteristic parameters for a description of such volatility. Also, theoretical models are generally established for availability matching, and a kind of optimal matching control model is established according to modern control theory, which provides a basic theoretical framework for future research in this field.

Chapter 7

Comparative Analysis on Discrete-Time Instantaneous Availability Models under Exponential Distribution

Discrete-time instantaneous availability models were specifically studied in the last chapter, which has some significance for the research on instantaneous availability fluctuation. However, as seen from the physical meaning of the discrete-time instantaneous availability models, their focus is to solve the problems in discrete-time systems. As we know, continuous-time problems can be approximately solved by discrete methods. Considering the advantages of discrete-time models in numerical calculation, this chapter comparatively analyzes the relationship between discrete-time distribution and continuous-time distribution, the relationship between numerical solutions of discrete-time models and analytical solutions of continuous-time models under exponential distribution, and the relationship between the numerical solutions of these two kinds of models. By measuring the possible errors in discrete-time models and exploring the feasibility of replacing the continuous-time models with discrete-time models, we lay a firm foundation for optimizing systems seeking instantaneous availability.

7.1 Analysis of Errors between Discrete-Time and Continuous-Time Distributions

Suppose the nonnegative random variable X follows the general continuous-time distribution, then the corresponding density function $f(t) \geq 0$ and distribution function are

$$F(t) = \int_0^t f(s)\,ds, \qquad t \geq 0. \tag{7.1}$$

Suppose T is the usual small sampling cycle. In actual engineering practices, the start of maintenance on faulted units and the normal operation of repaired units both happen at the sampling time points, which means the units have the discrete distribution $\tilde{X}$. The corresponding density function and distribution function are

$$f_T(k) = P\{kT \leq \tilde{X} < (k+1)T\} = \int_{kT}^{(k+1)T} f(t)\,dt, k = 0, 1, 2, \ldots \tag{7.2}$$

$$F_T(k) = \sum_{j=0}^{k} f_T(j), \ k = 0, 1, 2, \ldots, \tag{7.3}$$

then $\tilde{X}$ can be considered as a lattice distribution. Its distribution function can be defined as

$$F_T(t) = \sum_{j=0}^{k} f_T(j)kT \leq t < (k+1)T. \tag{7.4}$$

Unless otherwise specified, these two kinds of distributions are treated fairly. The following analyzes the difference between Formulas 7.1 and 7.2. Considering $\forall t, \exists k, s.t.,$ and $kT \leq t < (k+1)T$,

$$0 \leq F(t) - F_T(t) = \int_{kT}^{t} f(s)\,ds \leq \int_{kT}^{(k+1)T} f(s)\,ds \leq F((k+1)T) - F(kT)$$

$$\leq \max_{k \in N}[F((k+1)T) - F(kT)]\underline{\underline{\Delta}}H(T). \tag{7.5}$$

According to the definitions of density function and distribution function: $\int_0^{+\infty} f(t)\,dt = 1$, where $f(t)$ is the nonnegative function on $(0, +\infty)$. Obviously, there is $H(T) \xrightarrow{T \to 0} 0$; therefore, if any given precision requirement $\varepsilon_0 > 0$, there must be a sufficiently small T, enabling $H(T) \le \varepsilon_0$. At this time, it is somewhat reasonable to replace X with $\tilde{X}$ in the research.

Take the following Weibull distribution as an example. Suppose the nonnegative random variable X follows the continuous Weibull distribution of parameter (α, λ) and is called $W(\alpha, \lambda; t)$, where α is the shape parameter, and λ is the scale parameter. The corresponding density function and distribution parameter are

$$f(t) = \lambda\alpha(\lambda t)^{\alpha-1} e^{-(\lambda t)^{\alpha}}, \ t \ge 0; \quad \alpha, \lambda > 0$$

$$F(t) = 1 - e^{-(\lambda t)^{\alpha}}, \quad t \ge 0.$$

Supposing the sampling cycle is T, then the corresponding $\tilde{X}$ density function is

$$f_T(k) = F\big[(k+1)T\big] - F(kT) = e^{-(\lambda k T)^{\alpha}} - e^{-[\lambda(k+1)T]^{\alpha}}$$

$$= \left(e^{-(\lambda T)^{\alpha}}\right)^{k^{\alpha}} - \left(e^{-(\lambda T)^{\alpha}}\right)^{(k+1)^{\alpha}}, \quad k = 0, 1, 2, \dots.$$

$$F_T(t) = 1 - \left(e^{-(\lambda T)^{\alpha}}\right)^{(k+1)^{\alpha}}, \quad kT \le t < (k+1)T.$$

According to the definition of a discrete Weibull distribution, $\tilde{X}$ follows the discrete-time distribution where scale parameter is $e^{-(\lambda T)^{\alpha}}$ and the shape parameter is α. Assume $\alpha = 1.1$ and $\lambda = 0.1$, then

$$F(t) = 1 - e^{-(0.1t)^{1.1}}, \ t \ge 0$$

$$F_T(t) = 1 - \left(e^{-(0.1T)^{1.1}}\right)^{(k+1)^{1.1}}, \quad kT \le t < (k+1)T.$$

Therefore,

$$0 \le F_T(t) - F(t) \le \max\left\{e^{-(0.1kT)^{1.1}} - e^{(0.1(k+1)T)^{1.1}}\right\} \underline{\underline{\Delta}} H(T).$$

The upper bounds $H(T)$ for different sampling cycles are shown in Table 7.1. As the sampling cycle decreases, their difference will also gradually decrease, consistent with the results analyzed from Formula 7.4. Particular to this example, if the sampling cycle $T \le 0.1$, the difference will become even smaller. Setting the sampling cycle to 5, 1, and 0.5, then the corresponding discrete-time distribution functions and the distribution functions of the original continuous random variable are shown in Figures 7.1 and 7.2. Figure 7.2 is the enlarged view from 20-unit time to 30-unit time in Figure 5.1. As seen from the figures, when

Table 7.1 Difference between Distribution
Functions of Continuous-Time and Discrete-
Time Systems

T	H(T)
5	0.3722
1	0.0802
0.5	0.0403
0.1	0.0081
0.01	8.0768e−004

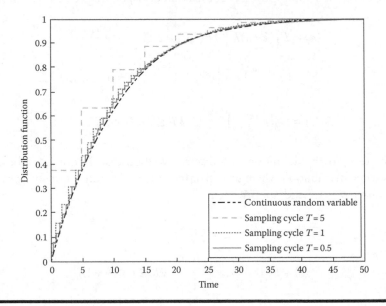

Figure 7.1 Comparison 1 among distribution functions of discrete random
variables corresponding to different sampling cycles and original continuous
random variables.

the sampling cycle T = 5, the corresponding discrete-time distribution functions
and the distribution functions of the original continuous random variable are
quite different from each other. As the sampling cycle decreases, such as when
T = 1 and T = 0.5, their difference will gradually decrease as well. When T = 0.5,
the corresponding discrete-time random variable has already approximately
approached the distribution function of the original continuous-time random
variable. At that moment, it is somewhat allowable to replace the original random
variable with the discrete random variable.

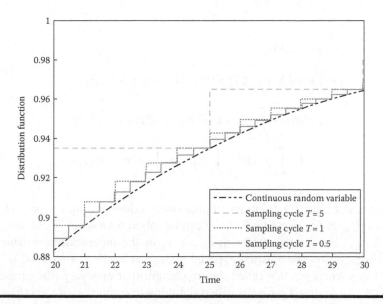

Figure 7.2 Comparison 2 among distribution functions of discrete random variables corresponding to different sampling cycles and original continuous random variables.

7.2 Comparative Analysis on Analytical Solutions and Numerical Solutions of Continuous-Time Models under Exponential Distribution

Suppose the system has only one unit, and its operation failure time X follows the general probability distribution $F(t)$. After the unit fails, it is repaired immediately. Its repair time Y follows the general probability distribution $G(t)$. After repair, the unit is immediately put into operation. Suppose the system is repaired as new, and X and Y are independent of each other. For simplicity, the unit at time 0 is usually supposed to be as new.

In order to differentiate among the different conditions of the system, its states are defined as

$$\begin{cases} Z(t) = 0 & \textit{system normal at } t \\ Z(t) = 1 & \textit{system failure at } t \end{cases} t \geq 0.$$

Reference [8] sets up the availability models for non-Markov single-unit repairable systems on the basis of renewal process

$$A(t) = R(t) + \int_0^t A(t-u)\, dQ\,(u) = R(t) + Q(t) * A(t),$$

where

$$R(t) = 1 - F(t)$$

$$Q(t) = P\{X + Y \leq t\} = F(t) * G(t) = \int_0^t q(u)\, du$$

$$A(t) = R(t) + \int_0^t A(t - u)\, dQ(u) = R(t) + Q(t) * A(t)$$

$$= 1 - \int_0^t f(s)\, ds + \int_0^t A(t - u) \int_0^u f(u - s)g(s)\, ds\, du.$$

With the numerical solution, approximate values u_1, u_2, ..., u_N of $u(t_1)$, $u(t_2)$, ..., $u(t_N)$ in position function $u(t)$ can be solved for a series of discrete points t_1, t_2, ..., t_N. The discrete values t_1, t_2, ..., t_N of the independent variable t are predefined; t_j is the general equally spaced node, that is, $t_1 = t_0 + h$, $t_2 = t_0 + 2h$, ..., $t_N = t_0 + Nh$, where $h > 0$ is called the step length that may vary when necessary; whereas u_1, u_2, ..., u_N is generally called the numerical solution of the initial value problem [76].

Setting $t_0 = 0$, and $t_1 = t_0 + h$, $t_2 = t_0 + 2h$, ..., $t_N = t_0 + Nh$, and considering that the integral equation still includes the integral terms, a rectangular formula is used herein:

$$A(t_j) = 1 - \int_0^{t_j} f(s)\, ds + \int_0^{t_j} A(t_j - u) \int_0^u f(u - s)g(s)\, ds\, du$$

$$\approx 1 - h\sum_{i=1}^{j} f(t_j) + h\sum_{i=1}^{j} A(t_j - t_i) \int_0^{t_i} f(t_i - s)g(s)\, ds$$

$$\approx 1 - h\sum_{i=1}^{j} f(jh) + h\sum_{i=1}^{j} \left\{ A[(j-i)h]h\sum_{l=1}^{i} f[(i-l)h]g(lh) \right\}$$

$$= 1 - h\sum_{i=1}^{j} f(jh) + h^2\sum_{i=1}^{j} \left\{ A[(j-i)h]\sum_{l=1}^{i} f[(i-l)h]g(lh) \right\}.$$

Setting

$$f_j = f(t_j), \quad g_j = g(t_j), \quad j = 1, 2, ...,$$

the numerical solution of instantaneous availability is

$$A_j = A(t_j), \quad j = 1, 2, ...$$

The iteration form of approximate instantaneous availability is

$$A_j = 1 - h\sum_{i=1}^{j} f_i + h^2 \sum_{i=1}^{j}\left(A_{j-i}\sum_{l=1}^{i} f_{i-l}g_l \right) \quad j = 1, 2, \ldots \quad (7.6)$$

Considering the solvability for the analytical form of accurate instantaneous availability and convenience for comparison, the system failure time and repair time after failure selected here both follow the exponential distribution form, which means the failure rates λ and μ are both constants. At this time, the actual instantaneous availability of the system is

$$A(t) = \frac{\mu}{\lambda+\mu} + \frac{\lambda}{\lambda+\mu}\exp\left[-(\lambda+\mu)t\right], \quad t \geq 0.$$

According to the iteration form (7.5), we draw up the approximate instantaneous availability curves of systems with corresponding step lengths of 0.02, 0.05, and 0.1 and the instantaneous availability curve of the original system, as shown in Figure 7.3. For clarity, a comparison figure is shown in Figure 7.4. As seen from the following two figures, the shorter the step length is, the smaller the error.

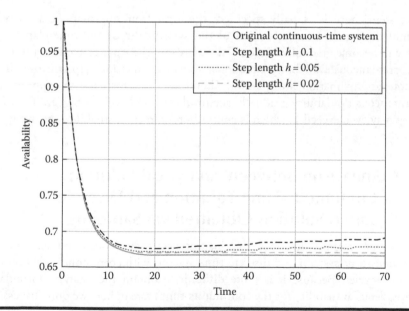

Figure 7.3 **Comparison 1 among approximate instantaneous availability curves with corresponding different step lengths and actual instantaneous availability curves.**

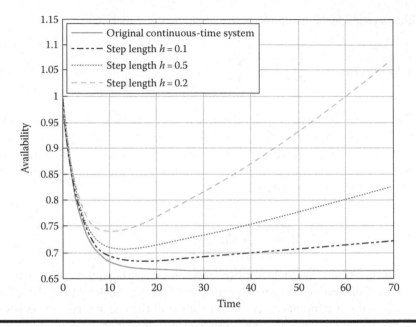

Figure 7.4 Comparison 2 among approximate instantaneous availability curves with corresponding different step lengths and actual instantaneous availability curves.

If the step length is casually specified, the error from approximate availability is slight during the early operation of the system. But as the time elapses, the error will become gradually bigger. This is because there are convolution terms in the instantaneous availability equations, and every iteration step will use all the previous approximation information. Therefore, the error from the approximate instantaneous availability gradually accumulates and becomes bigger. The availability may even exceed 1, which is contradictory to the availability meaning.

7.3 Comparison between Analytical Solution of Continuous-Time Equation and Discrete Equation Solution Obtained via Sampling

This section mainly describes the instantaneous availability models for the continuous-time systems and discrete-time systems with correspondingly different sampling cycles. Because it is quite difficult to obtain the analytic formula of instantaneous availability for the continuous-time system, here we only discuss the system under exponential distribution.

Suppose the system has only one unit, and its operation failure time X follows the general probability distribution $F(t)$. After the unit (system) fails, it is repaired immediately. Its repair time Y follows the general probability distribution $G(t)$. After repair, the unit is immediately put into operation. Suppose the system is repaired as new, and X and Y are independent of each other. For simplicity, the unit at time 0 is usually supposed to be as new.

Using the solvability for the analytical form of accurate instantaneous availability and convenience for comparison, the system failure time and repair time after failure selected here both follow the exponential distribution form, which means the failure rates λ and μ are both constants. At this time, the density function and distribution function of the failure time X are

$$f(t) = \lambda e^{-\lambda t}, \quad t \le 0$$

$$F(t) = \int_0^t f(s)\,ds = 1 - e^{-\lambda t}.$$

When the sampling cycle is T, the density function and distribution function of discrete failure time X_T from Formula 5.2 are

$$f_T(k) = e^{-\lambda kT}(1 - e^{-\lambda T})$$

$$F_T(t) = 1 - e^{-\lambda(k+1)T}, \quad k = 0, 1, 2, \dots$$

The discrete failure time X_T follows the geometric distribution, and then the system failure rate function is

$$\lambda(k) = 1 - e^{-\lambda T}, \quad k = 0, 1, 2, \dots$$

Therefore, it can be noted that the discrete repair time Y_T also follows the geometric distribution, and its density function and distribution function are

$$g_T(k) = e^{-\mu kT}(1 - e^{-\mu T})$$

$$G_T(t) = 1 - e^{-\mu(k+1)T}, \quad k = 0, 1, 2, \dots$$

The system repair rate function is

$$\mu(k) = 1 - e^{-\mu T}, \quad k = 0, 1, 2, \dots$$

Reference [8] sets up the instantaneous availability models for the original continuous-time systems with the Markov process:

$$A(t) = \frac{\mu}{\lambda + \mu} + \frac{\lambda}{\lambda + \mu} \exp[-(\lambda + \mu)t].$$

The state transition equations for the discrete-time system obtained from Chapter 4 are

$$\begin{cases} P_0(k+1, j+1) = P_0(k, j)(1 - \lambda(j)) & j \le k \\ P_1(k+1, j+1) = P_1(k, j)(1 - \mu(j)) & j \le k \end{cases} \tag{7.7}$$

$$\begin{cases} P_0(k+1, 0) = \sum_{j=0}^{k} \mu(j) P_1(k, j) \\ P_1(k+1, 0) = \sum_{j=0}^{T_0} \lambda(j) P_0(k, j) \end{cases} \tag{7.8}$$

with the additional definition:

$$P_0(k, j) = P_1(k, j) = 0, \quad j < k. \tag{7.9}$$

Generally, the system is supposed as new; that is,

$$(P_0(0, 0), P_1(0, 0), P_2(0, 0)) = (1, 0, 0). \tag{7.10}$$

The system instantaneous availability is

$$A(k) = \sum_{j=0}^{k} P_0(k, j), \tag{7.11}$$

where $P_0(k, j)$ is the probability that the system is already at 0 state (operating state of the system) at time point k for j unit time, and $P_1(k, j)$ is the probability that the system is already at 1 state (operating state of the system) at time point k for j unit time. For example, set $\lambda = 0.1$ and $\mu = 0.2$; the distribution functions of system failure time and repair time can be illustrated as Figure 7.5 and Figure 7.6.

As seen from Figure 7.7, when the sampling cycle decreases, the solution of the discrete-time equation gets closer and closer to the analytical solution, and the variation rule of the discrete-time equation solution is consistent with that of the analytical solution. Therefore, the solution of discrete-time equation can well replace the solution of continuous-time equation, especially the numerical solution.

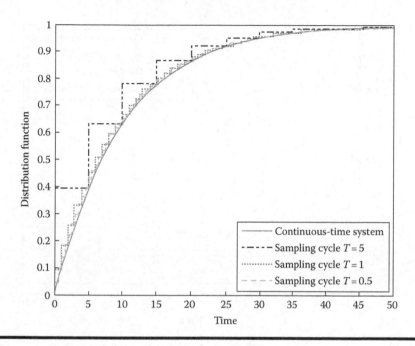

Figure 7.5   Distribution function of system failure time.

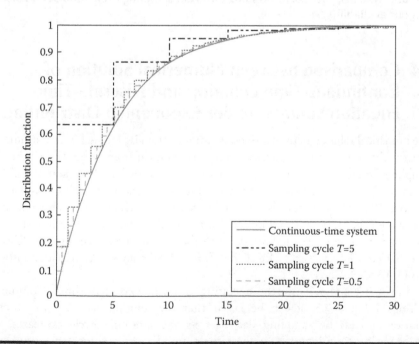

Figure 7.6   Repair time distribution after system failure.

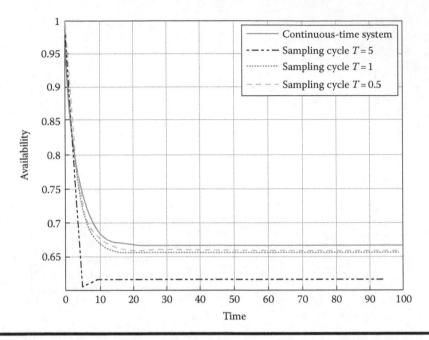

Figure 7.7 Comparison of instantaneous availability of discrete-time systems, one-unit systems, with corresponding different sampling cycles and actual instantaneous availability.

7.4 Comparison between Numerical Solution of Continuous-Time Equation and Discrete-Time Equation Solution under Exponential Distribution

The numerical solution of the continuous-time equation is obtained from Formula 7.5, whereas the discrete-time equation solution is obtained from Formula 7.10 through sampling. In order to see the errors produced by these two methods and considering the solvability for the analytical form of accurate instantaneous availability similarly to Sections 7.2 and 7.3, the selected system failure time and repair time after failure both follow the exponential distribution, which means the failure rates λ and μ are both constants. Assume $\lambda = 0.1$ and $\mu = 0.2$ with the experimental environment: PC computer, Celeron(R) 1.80 GHz CPU; 0.99 GB RAM; Windows XP; simulation software MATLAB® 7.3.

Tables 7.2 and 7.3 give the operating times of two algorithms. Combining Sections 7.2 and 7.3, it can be noted that the solution accuracy for discrete iteration process by sampling decreases as the sampling cycle decreases, but the operating time increases at the same time. This means that the accuracy is improved with more operating time as does the accuracy of numerical solutions

Table 7.2   CPU Time Consumption for Numerical Solutions Corresponding to Different Step Lengths

Step length h	0.02	0.05	0.1
CPU time (Unit: s)	535.6875	37	4.2813

Table 7.3   CPU Time Consumption for Discrete-Time Solutions Corresponding to Different Step Lengths

Sampling cycle T	0.1	0.2	0.5
CPU time (unit: s)	24.7969	2.9219	0.1250

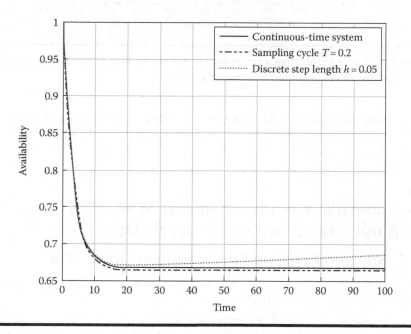

Figure 7.8   Comparison 1 between approximate availabilities obtained by sampling and discretization of continuous-time equations.

of continuous-time equations. We compare the approximate instantaneous availability corresponding to step length $h = 0.05$ and the instantaneous availability of an approximate discrete-time system corresponding to sampling cycle $T = 0, 2$, as shown in Figures 7.8 and 7.9. Figure 7.9 shows the availability variation from a 10 to 100 time interval. As seen from the figures that follow, the accuracy obtained from

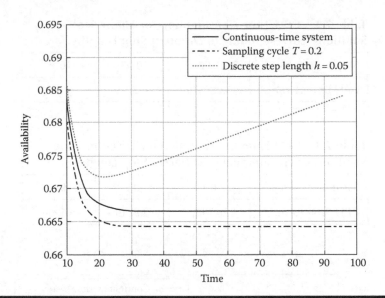

Figure 7.9 **Comparison 2 between approximate availabilities obtained by sampling and discretization of continuous-time equations.**

the sampling is obviously higher than that from discretization of continuous-time equations. But the operating time of sampling is just 2.9212 s, apparently much smaller than 37 s of discretization.

7.5 Comparison of Instantaneous Availability Models for Systems with Repair Delay

7.5.1 Numerical Method for Continuous-Time System

Suppose the system has only one unit, and its failure time X follows the general probability distribution $F(t)$. After the unit (system) has failed, it is repaired immediately. Its repair time Y follows the general probability distribution $G(t)$. After repair, the unit is immediately put into operation. Considering it has a repair delay, suppose the repair delay time W follows the general probability distribution $W(t)$. Suppose the system is repaired as new, and X and Y are independent of each other. For simplicity, the unit at time 0 is usually supposed to be as new.

Reference [8] sets up the availability models for non-Markov single-unit repairable systems on the basis of the renewal process as well:

$$A(t) = R(t) + \int_0^t A(t-u)\,dQ(u) = R(t) + Q(t) * A(t),$$

where

$$R(t) = 1 - F(t)$$

$$Q(t) = P\{X + Y + W \le t\} = F(t) * G(t) * W(t) = \int_0^t q(u)\, du$$

$$A(t) = R(t) + \int_0^t A(t-u)\, dQ(u) = R(t) + Q(t) * A(t)$$

$$= 1 - \int_0^t f(s)\, ds + \int_0^t A(t-u) \int_0^u w(u-s) \int_0^s f(s-v) g(v)\, dv\, ds\, du.$$

Next, we set $t_0 = 0$, and $t_1 = t_0 + h$, $t_2 = t_0 + 2h$, ..., $t_N = t_0 + Nh$, where $h > 0$ is called the step length. Considering the kernel of the integral equation still includes the integral terms, a rectangular formula is used herein:

$$A(t_j) = 1 - \int_0^{t_j} f(s)\, ds + \int_0^{t_j} A(t_j - u) \int_0^u w(u-s) \int_0^s f(s-v) g(v)\, dv\, ds\, du$$

$$\approx 1 - h \sum_{i=1}^{j} f(t_i) + h \sum_{i=1}^{j} A(t_j - t_i) \int_0^{t_i} w(t_i - s) \int_0^s f(s-v) g(v)\, dv\, ds$$

$$\approx 1 - h \sum_{i=1}^{j} f(t_i) + h \sum_{i=1}^{j} A(t_j - t_i) h \sum_{l=1}^{i} w(t_i - t_l) \int_0^{t_l} f(t_l - v) g(v)\, dv$$

$$\approx 1 - h \sum_{i=1}^{j} f(ih) + h \sum_{i=1}^{j} \left\{ A[(j-i)h] h \sum_{l=1}^{i} \left\{ w[(i-l)h] h \sum_{u=1}^{l} f[(l-u)h] g(uh) \right\} \right\}$$

$$= 1 - h \sum_{i=1}^{j} f(ih) + h^3 \sum_{i=1}^{j} \left\{ A[(j-i)h] \sum_{l=1}^{i} \left\{ w[(i-l)h] \sum_{u=1}^{l} f[(l-u)h] g(uh) \right\} \right\}.$$

Set

$$f_j = f(t_j),\ g_j = g(t_j),\ w_j = w(t_j), \quad j = 1, 2, \ldots$$

The numerical solution of instantaneous availability is

$$A_j = A(t_j), \quad j = 1, 2, \ldots$$

The iteration form of approximate instantaneous availability is

$$A_j = 1 - h \sum_{i=1}^{j} f_i + h^3 \sum_{i=1}^{j} \left\{ A_{j-i} \sum_{l=1}^{i} \left[w_{i-l} \sum_{u=1}^{l} (f_{l-u} g_u) \right] \right\}.$$

7.5.2 Description of Discrete-Time Models

Suppose the system has only one unit, and its failure time X follows the general probability distribution $F(t)$. After the system fails, it is repaired immediately. Its repair time is Y. Considering it has a repair delay, suppose the repair delay time is W, where X, Y, and W are independent of each other, and all follow the general discrete distribution.

In order to differentiate among the different conditions of the system, its states are defined as

$$\begin{cases} Z(t) = 0 & \textit{System normal at time point } k \\ Z(t) = 1 & \textit{System repaired after failure at time point } k \quad k = 0, 1, 2, \dots \\ Z(t) = 2 & \textit{System being repaired at time point } k \end{cases}$$

The system failure rate function and repair rate function are $\lambda(k)$ and $\mu(k)$, where $k = 0, 1, 2, \dots$. Because there is a repair delay, we define

$$\rho(k) = P\{W = k \mid W \geq k\} \quad k = 0, 1, 2, \dots$$

where $P_0(k, j)$ is the probability that the system is already at 0 state (operating state of the system) at time point k for j unit time; $P_1(k, j)$ is the probability that the system is already at 1 state (operating state of the system) at time point k for j unit time; and $P_2(k, j)$ is the probability that the system is already at 2 state (operating state of the system) at time point k for j unit time, namely $j = 0, 1, \dots, k - 1$:

$$\begin{cases} P_0(k, j) = P\{Z(k) = \dots = Z(k - j) = 0, Z(k - j - 1) = 1\} \\ P_1(k, j) = P\{Z(k) = \dots = Z(k - j) = 1, Z(k - j - 1) = 2\} \\ P_2(k, j) = P\{Z(k) = \dots = Z(k - j) = 2, Z(k - j - 1) = 0\}. \end{cases}$$

When $j = k$,

$$\begin{cases} P_0(k, k) = P\{Z(k) = \dots = Z(0) = 0\} \\ P_1(k, k) = P\{Z(k) = \dots = Z(0) = 1\} \\ P_2(k, k) = P\{Z(k) = \dots = Z(0) = 2\}. \end{cases}$$

When $k < j$,

$$P_0(k, j) = 0, P_1(k, j) = 0, P_2(k, j) = 0.$$

The mathematical model for this repairable system can be obtained through probability analysis:

$$\begin{cases} P(k+1, j+1) = H(j)P(k, j) \\ P(k+1, 0) = \displaystyle\sum_{j=0}^{k} B(j)P(k, j), \end{cases}$$

where

$$P(k, j) = (P_0(k, j), P_1(k, j), P_2(k, j))^T$$

$$H(j) = \begin{bmatrix} 1-\lambda(j) & 0 & 0 \\ 0 & 1-\mu(j) & 0 \\ 0 & 0 & 1-\rho(j) \end{bmatrix}$$

$$B(j) = \begin{pmatrix} 0 & \mu(j) & 0 \\ 0 & 0 & \rho(j) \\ \lambda(j) & 0 & 0 \end{pmatrix}.$$

Generally, the new system is supposed as available; that is,

$$P_0(0,0) = 1, P_1(0,0) = 0, P_2(0,0) = 0.$$

The system instantaneous availability is

$$A(k) = \sum_{j=0}^{k} P_0(k, j).$$

7.5.3 Model Comparison

Considering the solvability for the analytical form of accurate instantaneous availability and convenience for comparison, the system failure time, repair time after failure, and repair delay time selected here all follow the exponential distribution as parameters λ, μ, and ρ. According to Reference [8], this system is a Markov repairable system, and its state is defined as follows:

$$\begin{cases} Z(t) = 0 & \textit{System normal at time point } t \\ Z(t) = 1 & \textit{System repaired after failure time point } t \\ Z(t) = 2 & \textit{System being repaired at time point } t \end{cases}$$

Suppose

$$P_j(t) = P\{Z(t) = j\}, \quad j = 0, 1, 2.$$

Then

$$\begin{pmatrix} P_0(t) \\ P_1(t) \\ P_2(t) \end{pmatrix}' = \begin{bmatrix} -\lambda & 0 & \mu \\ \lambda & -\rho & 0 \\ 0 & \rho & -\mu \end{bmatrix} \begin{pmatrix} P_0(t) \\ P_1(t) \\ P_2(t) \end{pmatrix}$$

$$(P_0(t), P_1(t), P_2(t))\,|_{t=0} = (1, 0, 0).$$

The system instantaneous availability is

$$A(t) = P_0(t).$$

For example, set $\lambda = 0.2$, $\mu = 2$, and $\rho = 3.6$:

$$A(t) = \frac{5}{13} e^{-2.6t} + \frac{1}{4} e^{-3.2t} + \frac{45}{52}, \quad t \geq 0.$$

Tables 7.4 and 7.5 give the operating times of two algorithms. It can be noted that the solution accuracy for discrete iteration process by sampling decreases as the sampling cycle decreases, but the operating time increases at the same time.

Table 7.4   CPU Time Consumption for Numerical Solutions Corresponding to Different Step Lengths

Step length h	0.05	0.02
CPU time (unit: s)	2.3438	74.9219

Table 7.5   CPU Time Consumption for Discrete-Time Solutions Corresponding to Different Step Lengths

Sampling cycle T	0.05	0.01
CPU time (unit: s)	0.2500	33.6406

This means the accuracy is improved with more operating time, as does the accuracy of numerical solutions of continuous-time equations. Through the comparison between the approximate instantaneous availability corresponding to step length $h = 0.02$ and the calculation result in case of sampling cycle $T = 0.05$ (as seen from Figures 7.10 and 7.11), the accuracy obtained from the sampling is obviously

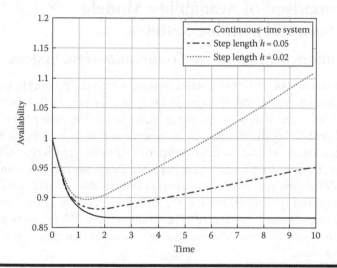

Figure 7.10 **Comparison of approximate instantaneous availability curves with corresponding different step lengths and actual instantaneous availability curves.**

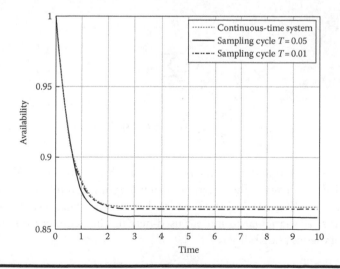

Figure 7.11 **Comparison of instantaneous availability of discrete-time systems, one-unit systems with repair delay, with corresponding different sampling cycles and actual instantaneous availability.**

higher than that from numerical solution of continuous-time equations. But the operating time of sampling is just 0.2500 s, apparently much smaller than 74.9218 s of numerical solution of continuous-time equations.

7.6 Comparison of Availability Models for Series Repairable Systems

7.6.1 Numerical Method for Continuous-Time System

The system consists of n units in a series, among which the failure time of ith unit X_i follows the exponential distribution $F_i(t) = 1 - e^{-\lambda_i t}$, where $t \geq 0$. After the system fails, it is fixed immediately by the repair equipment. The repair time Y_i follows the general probability distribution $G_i(t)$, where $i = 1, 2, ..., n$. Suppose the failure time distribution of the system after unit repair is consistent with that of new units, namely it is still called $F_i(t)$, and n units operate independently from one another. The unit failure time and repair time are also independent of each other. When a unit has failed and is being repaired, other units stop running so there are no more failures. At that moment, the system is in a failure state. When the faulted unit is repaired, n units are immediately put into operation. At that time, the system is in an operating state.

The instantaneous availability of the system $A(t)$ meets

$$A(t) = e^{-\Lambda t} + Q(t)^*A(t),$$

where

$$\Lambda = \sum_{j=1}^{n} \lambda_i$$

$$Q(t) = \sum_{j=1}^{n} \lambda_j \int_0^t G_j(t-u) \exp(-\Lambda u)\, du$$

We can set $t_0 = 0$ and $t_1 = t_0 + h$, $t_2 = t_0 + 2h$, ..., $t_N = t_0 + Nh$, ..., where $h > 0$ is called step length. The selection of step length h can follow the conditions $\exists\, T_0 \in N$, $T = T_0 h$. Considering that the kernel of integral equation still includes the integral terms, a rectangular formula is used herein. The iteration form of approximate instantaneous availability is

$$A_j = \exp\left[-(j\Lambda h)\right] + h^2 \sum_{i=1}^{j}\left(A_{j-i}\sum_{u=1}^{n}\lambda_u\left(\sum_{v=1}^{i} g_u(i-v)\exp[-(v\Lambda h)]\right)\right),$$

where A_j is the approximate instantaneous availability in case $t_j = jh$ for the system instantaneous availability $A(t)$:

$$g_u[(i-v)h] = g_{u(i-v)}.$$

7.6.2 Discrete Model

Suppose the system consists of n units in a series, among which the failure time of ith unit X_i follows the geometric distribution, whereas the repair time after failure Y_i follows the general probability distribution. The failure rate function and repair rate function of every unit are $\lambda(k) = \lambda$ and $\mu(k)$, respectively. Generally, in the discrete-time repairable system, the failure can occur in multiple units at the same time. Because this kind of probability is small, it can be ignored, especially in the case of a very small sampling cycle. This is also a proper assumption. Once failure occurs in a unit, the system malfunctions, and other units stop running so there are no more failures. After the faulty unit is repaired, n units are immediately put into operation. At this time, the system is in an operating state, as shown in Figure 7.12.

In order to differentiate the different conditions of the system, define

$$Z(k) = \begin{cases} 0 & \text{Units normal at time point } k \\ i & \text{Unit } i \text{ repaired after failure at time point } k \quad i = 1, 2, ..., n, \end{cases}$$

where $P_0(k)$ is the probability that the system is already at 0 state at time point k for j unit time; $P_i(k, j)$ is the probability that the system is already at i state at time point k within j time, namely $j = 0, 1, ..., k - 1$:

$$\begin{cases} P_0(k) = P\{Z(k) = 0\} \\ P_i(k, j) = P\{Z(k) = Z(k-1) = ... = Z(k-j) = i, Z(k-j-1) = 0\}. \end{cases}$$

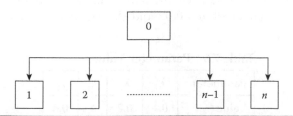

Figure 7.12 N-unit series system model.

Generally, the new system is supposed to be available, namely $P_0(0) = 1$. When $k \le j$, $P_i(k, j) = 0$, where $i = 1, 2, ..., n$.

The following can be obtained through probability analysis:

$$
\begin{cases}
P_0(k+1) = P_0(k)\prod_{i=1}^{n}(1-\lambda_i) + \sum_{i=1}^{n}\sum_{j=1}^{k}P_i(k,j)\mu_i(j) \\[2ex]
P_i(k+1, j+1) = P_i(k,j)(1-\mu_i(j)) \\[2ex]
P_i(k+1,0) = \lambda_i P_0(k) \qquad\qquad i = 1, 2, ..., n.
\end{cases}
$$

The system instantaneous availability at time point k is

$$
A(k) = P_0(k).
$$

7.6.3 Model Comparison

Now we will use a two-unit series system as an example. Considering the solvability of the analytical form of accurate instantaneous availability and the convenience for comparison, the failure time and repair time after failure selected here for the unit $i(i = 1, 2)$ both follow the exponential distribution as parameters λ_i and μ_i. Parameter values obtained are shown in Table 7.6.

According to Reference [8], the system instantaneous availability is

$$
A(t) = \frac{20}{27} + \frac{2}{9}e^{-1.2t} + \frac{1}{27}e^{-0.9t} \quad t \ge 0.
$$

Tables 7.7 and 7.8 give the operating times of two algorithms. It can be noted that the solution accuracy for the discrete iteration process by sampling decreases as the sampling cycle decreases, but the operating time increases at the same time. This means the accuracy is improved with more operating time, as does the accuracy of numerical solutions of continuous-time equations. Through the comparison between the approximate instantaneous availability corresponding to step length $h = 0.02$ and the calculation result in the case of

Table 7.6 Parameter Values

Parameter	λ_1	λ_2	λ_1	λ_2
Value	0.1	0.2	1	0.8

Table 7.7 CPU Time Consumption for Numerical Solutions Corresponding to Different Step Lengths

Step length h	0.05	0.02
CPU time (unit: s)	0.4375	4.7813

Table 7.8 CPU Time Consumption for Discrete-Time Solutions Corresponding to Different Step Lengths

Sampling cycle T	0.5	0.1	0.05
CPU time (unit: s)	0.1250	0.1563	0.2500

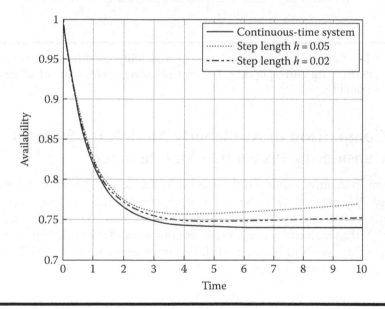

Figure 7.13 Comparison of approximate instantaneous availability curves with corresponding different step lengths and actual instantaneous availability curves.

sampling cycle T = 0.05 (as seen in Figures 7.13 and 7.14), the accuracy obtained from the sampling is obviously higher than that from the numerical solution of continuous-time equations. But the operating time of sampling is just 0.2500 s, apparently much smaller than 4.7813 s—the numerical solution of continuous-time equations.

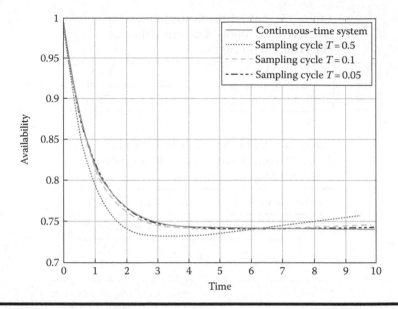

Figure 7.14 **Comparison of instantaneous availability of discrete-time systems, series systems, with corresponding different sampling cycles and actual instantaneous availability.**

7.7 Comparison of Availability Models for Systems Considering Preventive Maintenance

The term "preventive maintenance" refers to when the equipment operates to the specified time interval T, it also needs preventive maintenance even though no failures occur. If the equipment has failures before reaching the specified time interval T, then maintenance needs to be conducted, and the operating time of the equipment also needs to be re-recorded; this is where the time interval T is called the preventive maintenance cycle.

7.7.1 Numerical Solution of Continuous-Time Model

Suppose the unit failure time X follows the general probability distribution $F(t)$. When the unit still has no failures occurring as it runs to the specified time T, then the unit is maintained preventively. The repair time Y_p of preventive maintenance follows the general probability analysis $G_p(t)$. If the unit has a failure occurring before reaching the specified time T, then the repairable maintenance immediately takes place. The repair time Y_f follows the general probability distribution $G_f(t)$. After the unit is maintained preventively or repaired, it is recovered as new. Now suppose that X, Y_p, and Y_f are independent of each other.

Reference [8] solves the system instantaneous availability through the renewal process

$$A(t) = P\left\{\tilde{X} > t\right\} + P\left\{\tilde{X} + \tilde{Y} \le t\right\} * A(t),$$

where

$$\tilde{X} = \begin{cases} X & \text{when } X \le T \\ T & \text{when } X > T \end{cases}, \quad \tilde{Y} = \begin{cases} Y_f & \text{when } X \le T \\ Y_P & \text{when } X > T \end{cases}$$

$$P\left\{\tilde{X} > t\right\} = \begin{cases} 1 - F(t) \, T & \text{when } t < T \\ 0 & \text{when } t \ge T \end{cases}$$

$$
\begin{aligned}
P\left\{\tilde{X} + \tilde{Y} \le t\right\} &= P\left\{\tilde{X} + \tilde{Y} \le t, X \le T\right\} + P\left\{\tilde{X} + \tilde{Y} \le t, X > T\right\} \\
&= P\left\{X + Y_f \le t, X \le T\right\} + P\left\{Y_p \le t - T, X > T\right\} \\
&= \begin{cases} \displaystyle\int_0^t f(x) G_f(t-x)\, dx \\[2mm] \displaystyle\int_0^T f(x) G_f(t-x)\, dx + G_p(t-T)[1-F(T)]. \end{cases}
\end{aligned}
$$

Therefore,

$$
A(t) = \begin{cases} \displaystyle 1 - \int_0^t f(x)\, dx + \int_0^t A(t-u) q_1(u)\, du, & \text{when } t \le T \\[4mm] \displaystyle \int_0^T A(t-u) q_1(u)\, du + \int_T^t A(t-u) q_2(u)\, du, & \text{when } t \le T. \end{cases}
$$

where

$$q_1(u) = \int_0^u f(x) g_f(u-x)\, dx$$

$$q_2(u) = \int_0^T f(x) g_f(u-x) + g_p(u-T)[1-F(T)].$$

Then, we set $t_0 = 0$ and $t_1 = t_0 + h$, $t_2 = t_0 + 2h$, ..., $t_N = t_0 + Nh$, ... The selection of step length h can follow the conditions: $\exists T_0 \in N$, $T = T_0 h$. Considering that the kernel of integral equation still includes the integral terms, the rectangular formula is used herein. The iteration form of approximate instantaneous availability is

$$
A_j = \begin{cases} \left[1 - h\sum_{i=1}^{j} f_i + h^2 \sum_{i=1}^{j}\left(A_{j-i}\sum_{l=1}^{i} f_l g_{f(i-l)} \right) \right], & j \le T_0 \\[4mm] \left[h^2 \sum_{i=1}^{T_0}\left(A_{j-i}\sum_{l=1}^{i} f_l g_{f(i-l)} \right) + h\sum_{i=T_0+1}^{j}\left[A_{j-i}\left[\sum_{l=1}^{T_0} f_l g_{f(i-l)} + g_{p(i-T_0)}\left(1 - \sum_{l=1}^{T_0} f_l \right) \right] \right] \right], & j > T_0 \end{cases}
$$

where A_j is the approximate instantaneous availability in case $A(t)$ for the system instantaneous availability $t_j = jh$:

$$
f_j = f(t_j), \quad g_{f(j)} = g_f(t_j), \quad g_{p(j)} = g_p(t_j), \quad j = 1, 2, \ldots
$$

7.7.2 Discrete-Time Model

The system availability model can be obtained from section 4.4 as follows:

$$
A(k) = \sum_{j=0}^{k} P_0(k, j) = \sum_{j=0}^{T_0} P_0(k, j).
$$

7.7.3 Model Comparison

Although there is not much sense in conducting preventive maintenance when there is a constant failure rate, we can still easily describe the problem by supposing the system failure rate $\lambda(t) = 0.1$, repair rate of repairable maintenance $\mu_1(t) = 0.6$, the repair rate of preventive maintenance is $\mu_2(t) = 1$, and the prevent maintenance cycle of system $T = 5$.

Considering that the instantaneous availability models with preventive maintenance have no analytical solutions, two kinds of numerical solutions will be compared. Figures 7.15 and 7.16 give the variation rules of instantaneous availability calculated using two kinds of algorithms. As seen from these figures, when the numerical solution selects enough step length, the calculation accuracy can meet corresponding standards as does the sampling cycle. Tables 7.9 and 7.10 give the operating times of the two algorithms. As seen from the comparison between the approximate instantaneous availability corresponding to selected step lengths $h = 0.01$ and 0.005 and the calculation results of selected sampling cycles $T = 0.1$ and 0.005, the calculation accuracies and variation rules are basically same under these two methods. Obviously, the calculation time through sampling is less than

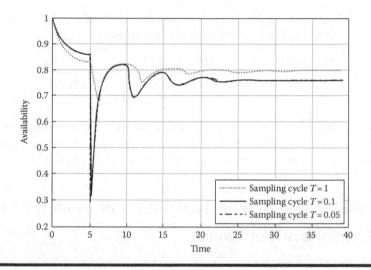

Figure 7.15 **Comparison of instantaneous availability of discrete-time systems with corresponding different sampling cycles.**

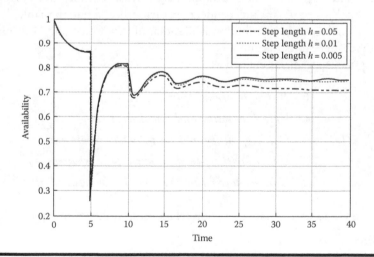

Figure 7.16 **Comparison of instantaneous availability curves with different step lengths.**

Table 7.9 **CPU Time Consumption for Discrete-Time Solutions Corresponding to Different Step Lengths**

Sampling cycle T	1	0.1	0.05
CPU time (unit: s)	0.0781	0.3750	16.3438

Table 7.10 CPU Time Consumption for Numerical Solutions Corresponding to Different Step Lengths

Step length h	0.05	0.01	0.005
CPU time (unit: s)	1.3906	97.7500	799.0156

that of numerical solution of continuous-time equations. The operating times from sampling are 0.3750 s and 16.3438 s, whereas the operating times from numerical solution of continuous-time equations are 97.7500 s and 799.0156 s.

7.8 Parallel System with Two Different Models of Units

7.8.1 Continuous-Time Model

Consider the parallel system consisting of two different models of units and one repairing device, as shown in Figure 7.17. Suppose the failure time X_i of unit i follows the exponential distribution $F_i(t) = 1 - e^{-\lambda_i t}$, $t \geq 0$, $(\lambda_i > 0)$, and the repair time after failure Y_i follows the general probability distribution $G_i(t)$. Further suppose X_1, X_2, Y_1, and Y_2 are independent of each other. After the faulted unit is repaired, its failure time distribution is like that of new units, and both units are new as the system starts to run.

The system state $Z(t)$ is defined as

$$Z(t) = \begin{cases} 0 & \textit{Both units operating at t} \\ 1 & \textit{Unit 2 operating and unit 1 repair at t} \\ 2 & \textit{Unit 1 operating and unit 2 under repair at t} \\ 3 & \textit{Unit 1 under repair and unit 2 to be repaired at t} \\ 4 & \textit{Unit 2 under repair and unit 1 to be repaired at t.} \end{cases}$$

Therefore, the system state transition relationship is shown in Figure 7.18.

Define:

$$A_i(t) = P\{\textit{System normal at time point } P|Z(0) = i\}.$$

Because the system is supposed to be as new at the start, then the system instantaneous $A_0(t)$ can meet

$$\begin{cases} A_0(t) = Q_{01}(t) * A_1(t) + Q_{02}(t) * A_2(t) + \left[1 - Q_{01}(t) - Q_{02}(t)\right] \\ A_1(t) = Q_{10}(t) * A_0(t) + Q_{12}(t) * A_2(t) + \left[1 - Q_{10}(t) - Q_{13}(t)\right], \\ A_2(t) = Q_{20}(t) * A_0(t) + Q_{21}(t) * A_1(t) + \left[1 - Q_{20}(t) - Q_{24}(t)\right] \end{cases}$$

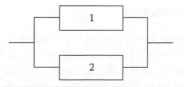

Figure 7.17 **Two-unit parallel system.**

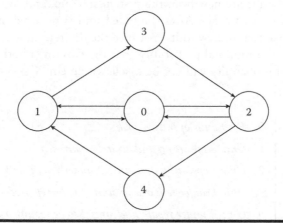

Figure 7.18 **System state transition relationship in continuous time.**

where

$$Q_{01}(t) = P\{X_1 \le t, X_1 < X_2\} = \frac{\lambda_1}{\lambda_1 + \lambda_2}\left[1 - e^{-(\lambda_1 + \lambda_2)t}\right]$$

$$Q_{02}(t) = P\{X_2 \le t, X_2 < X_1\} = \frac{\lambda_1}{\lambda_1 + \lambda_2}\left[1 - e^{-(\lambda_1 + \lambda_2)t}\right]$$

$$Q_{10}(t) = P\{Y_1 \le t, Y_1 < X_2\} = \int_0^t e^{-\lambda_2 u}\,dG_1(u)$$

$$Q_{20}(t) = P\{Y_2 \le t, Y_2 < X_1\} = \int_0^t e^{-\lambda_1 u}\,dG_2(u)$$

$$Q_{12}(t) = P\{Y_1 \le t, X_2 < Y_1\} = \int_0^t (1 - e^{-\lambda_2 u})\,dG_1(u)$$

$$Q_{21}(t) = P\{Y_1 \le t, X_1 < Y_2\} = \int_0^t \left(1 - e^{-\lambda_1 u}\right)dG_2(u)$$

$$Q_{13}(t) = P\{X_2 \le t, X_2 < Y_1\} = \int_0^t \left(1 - G_1(u)\right)d\left(1 - e^{-\lambda_2 u}\right)$$

$$Q_{24}(t) = P\{X_1 \le t, X_1 < Y_2\} = \int_0^t \left(1 - G_2(u)\right)d\left(1 - e^{-\lambda_1 u}\right).$$

7.8.2 Discrete-Time Model

Consider the parallel system consisting of two different models of units and one repairing device, as shown in Figure 7.17. Suppose the failure time X_i of unit i follows the geometric distribution, and the repair time after failure Y_i follows the general probability distribution. The failure rate function and repair rate function of unit i are $\lambda_i(k) = \lambda_i$ and $\mu_i(k)$, where $k = 0, 1, 2, \ldots$ and $i = 1, 2$. Further suppose the two units are new from the system start-up, and $X_1, X_2, Y_1,$ and Y_2 are independent of each other. After the failed unit is repaired, its failure time distribution is like that of new units. Because the discrete time interval is quite small, and there is a minimal probability that the two units both have failures, thus this kind of probability will not be discussed herein. The system state $Z(k)$ is defined as

$$Z(k) = \begin{cases} 0 & \textit{Both units operating at } k \\ 1 & \textit{Unit 2 operating and unit 1 repair at } k \\ 2 & \textit{Unit 1 operating and unit 2 under repair at } k \\ 3 & \textit{Unit 1 under repair and unit 2 to be repaired at } k \\ 4 & \textit{Unit 2 under repair and unit 1 to be repaired at } k. \end{cases}$$

The system state transition relationship in discrete time is shown in Figure 7.18.

Just like models in section 4.5, the system instantaneous availability at time point k can be written as

$$A(k) = P_0(k) + \sum_{j=0}^{k} P_1(k, j) + \sum_{j=0}^{k} P_2(k, j).$$

7.8.3 Model Comparison

Considering the solvability for the analytical form of instantaneous availability and convenience for comparison, the failure time and repair time after the failure of each unit in the system selected here both follow the exponential distribution form, which means the failure rates λ and μ are both constants. Suppose the two units are the same model. According to Reference [8], the instantaneous availability of this system is

$$A(t) = \frac{2\lambda\mu + \mu^2}{2\lambda^2 + 2\lambda\mu + \mu^2} - \frac{2\lambda^2 (s_2 e^{s_1 t} - s_1 e^{s_2 t})}{s_1 s_2 (s_1 - s_2)}.$$

And s_1, s_2 are two roots of the equation, where

$$\Delta(s) = s^2 + (3\lambda + 2\mu)s + (2\lambda^2 + 2\lambda\mu + \mu^2),$$

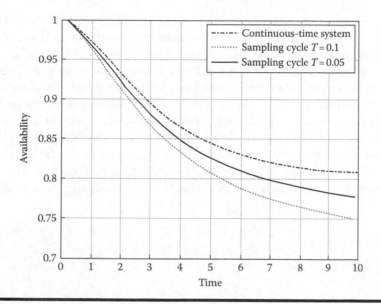

Figure 7.19 Comparison of instantaneous availability of discrete-time systems, one-unit systems with preventive maintenance, with corresponding different sampling cycles and actual instantaneous availability.

namely,

$$s_1, s_2 = \frac{1}{2}\left[-(3\lambda + 2\mu) \pm \sqrt{\lambda^2 + 4\lambda\mu}\right] < 0.$$

Setting

$$\lambda = 0.2, \quad \mu = 0.4,$$

the system instantaneous availability at this moment is

$$A(t) = \frac{4}{5} + \frac{1}{3}e^{-0.4t} - \frac{2}{15}e^{-t} \quad t \geq 0.$$

As shown in Figure 7.19.

7.9 Summary

This chapter studies the feasibility of replacing continuous-time models with instantaneous availability models for discrete-time repairable systems. The research concludes that when the discrete-time sampling interval is two hours, certain differences occur between the numerical solutions of these two kinds of models. Because the numerical solution of continuous-time model is used to solve the integral equations, with quite large errors, the solution variation rule often differs from the accurate analytical solution. As time elapses, these errors grow larger and larger. However, in cases where the sampling cycle is properly set, the solution accuracy and variation rule of the discrete-time model are almost completely consistent with the accurate analytical solution. From the view of calculation accuracy, the discrete-time model can replace the continuous-time model. Meanwhile, through the comparison, the calculation amount of discrete-time model is far less than the numerical solution of the continuous-time model; this gives actual meaning to the analysis of the instantaneous availability fluctuation with discrete-time models.

Chapter 8

Analysis of Instantaneous Availability Fluctuation under Truncated Discrete Weibull Distribution

8.1 Overview

Using the parameters describing the system instantaneous availability fluctuation proposed earlier, this chapter will discuss the fluctuation parameter characteristics without changing the mean time to repair, mean time between failures, or the mean logistic delay time of the system. Through simulation analysis, the variation rule of system availability fluctuation parameters corresponding to the characteristic parameters of related time distributions will be obtained to determine the design based on the instantaneous availability fluctuation.

Following Chapter 6, suppose the parameters describing instantaneous availability fluctuation are I, and the selected parameters mainly include availability amplitude, occurrence time of minimum availability, adaptation time, and mean availability (one or more), depending on different models; here, I refers to all the fluctuation parameters in a general sense.

8.2 Weibull Truncated and Discrete Distribution

In 1951, Weibull proposed the Weibull distribution for engineering demands [89]. Established practices show that this distribution is of great importance in reliability engineering. Weibull distribution has been widely used in many aspects, such as determining steel fatigue life, structural strength of glass, pipe corrosion, and so forth [90–95]. In recent years, References [96–99] further generalized and extended the Weibull distribution. Because Weibull distribution is strongly represented in reliability engineering, the following chapters will mainly follow a two-parameter discrete Weibull distribution.

In general reliability design, the probability distribution is usually the theoretical distribution because theoretical calculation is easy to undertake. However, this is somewhat different from reality because different variables proceed toward infinity in real engineering [100], especially when considering the service life or repair time of units, for example. In order to more accurately describe real problems, various kinds of truncated distributions are widely used in reliability engineering. In this chapter, the general discrete Weibull distribution is properly modified for setting up one-sided, truncated, and discrete Weibull distribution models. There are two models for the one-sided, truncated, and discrete Weibull distribution models:

1. Set the nonnegative random variable X, and the distribution is

$$p_k = \Pr\{X = k\} = \frac{\left(\alpha^{k^\beta} - \alpha^{(k+1)^\beta}\right)}{KK}, \quad k = 0, 1, 2, \ldots, n-1$$

$$p_n = \Pr\{X = k\} = \alpha^{n^\beta} \tag{8.1}$$

$0 < \alpha < 1, \beta > 0,$
which follows the truncated and discrete Weibull distribution with scale parameter α and shape parameter β, where n is the truncated point, and $KK = 1 - q^{n^\beta}$.

2. Set the nonnegative random variable X, and the distribution is

$$p_k = \Pr\{X = k\} = \alpha^{k^\beta} - \alpha^{(k+1)^\beta}, \quad k = 0, 1, 2, \ldots, n-1$$

$$p_n = \Pr\{X = k\} = \alpha^{n^\beta} \tag{8.2}$$

$0 < \alpha < 1, \beta > 0.$

If X is the system service life, then its failure rate is

$$\lambda(k) = 1 - \alpha^{(k+1)^{\beta} - k^{\beta}}, \quad k = 0, 1, 2, \dots, n-1 \tag{8.3}$$

$$\lambda(n) = 1,$$

where X follows the truncated and discrete Weibull distribution with scale parameter α and shape parameter β, called Weibull(α, β), where n is the truncated point.

If one compares these two models, it can be noted that as long as n is big enough to make $KK \approx 1$, the two models may not vary much from each other. Considering the simplicity of failure rate, this chapter will uses the second truncated model. The truncated point needs to be big enough to minimize the differences within the truncated distribution original probability distribution. Unless otherwise specified, the truncated point of the truncated and discrete Weibull distribution herein is selected as the upper boundary of the entire research time period.

8.3 Analysis of Instantaneous Availability Fluctuation for General Repairable System

For a general repairable system, the system instantaneous availability $A(k)$ meets Formulas 5.1 through 5.3. Suppose the service life and repair time after service of the units follow the probability distributions Weibull(α_1, β_1) and Weibull(α_2, β_2), respectively. Then,

$$I = I(A(.)) = I(\lambda(.), \mu(.)) = I(\alpha_1, \beta_1, \alpha_2, \beta_2) \tag{8.4}$$

where I is a four-variable function, where three parameters was chosen to depict the availability fluctuation: availability amplitude M, occurrence time of minimum availability T_0, and adaptation time T.

Two sets are divided in the numerical simulation: the probability distribution Weibull(α_2, β_2) of fixed repair time and mean time between failure $MTBF$ and the fixed service life distribution Weibull(α_1, β_1) and mean time to repair $MTTR$.

8.3.1 Effects of Scale and Shape Parameters in Service Life Distribution on Instantaneous Availability Fluctuation

Shaping parameter β_2 and scale parameter α_2 in fixed repair time distribution and mean time between failure $MTBF$: Set $\beta_2 = 3$ and 0.995, then $MTTR = 5.7178$.

Fixing a value for *MTBF* and then dividing the simulation into three sets as *MTBF* varies as follows:

1. Suppose *MTBF* = 9.4283, then the inherent availability A = 0.6225. Match α_1 and β_1, then obtain the fluctuation characteristic parameter values for the system, as shown in Table 8.1.
2. Suppose *MTBF* = 28.5179, then the inherent availability A = 0.8331. Match α_1 and β_1, then obtain the fluctuation characteristic parameter values for the system, as shown in Table 8.2.
3. Suppose *MTBF* = 40.1283, then the inherent availability A = 0.8754. Match α_1 and β_1, then obtain the fluctuation characteristic parameter values for the system, as shown in Table 8.3.

The data results for Tables 8.1 through 8.3 in graphs are shown in Figures 8.1 through 8.3. The x-coordinate is the shape parameter β_1 of service life distribution, and the y-coordinate is some fluctuation characteristic parameters.

Table 8.1 System Fluctuation Characteristic Parameters Corresponding to Different (α_1, β_1) Combinations When *MTBF* = 9.4283

| Experiment | Parameter | | | | |
	α_1	β_1	T_0	M	T
1	0.6	0.478	4	0.2043	120
2	0.7	0.589	5	0.1384	59
3	0.8	0.746	5	0.0811	27
4	0.9	1.023	7	0.0297	10
5	0.91	1.066	7	0.0265	10
6	0.93	1.169	8	0.0208	11
7	0.95	1.308	9	0.0207	12
8	0.97	1.522	10	0.0322	15
9	0.99	1.991	11	0.0782	31
10	0.993	2.145	11	0.0954	32
11	0.996	2.389	12	0.1238	33
12	0.999	3.000	12	0.1999	62
13	0.9993	3.158	12	0.2187	63
14	0.9997	3.535	12	0.2613	77

(Continued)

Table 8.1 *(Continued)*   System Fluctuation Characteristic Parameters Corresponding to Different (α_1, β_1) Combinations When *MTBF* = 9.4283

Experiment	Parameter				
	α_1	β_1	T_0	M	T
15	0.99973	3.582	12	0.2664	77
16	0.99976	3.635	12	0.2721	77
17	0.99979	3.695	12	0.2784	77
18	0.99983	3.789	12	0.2881	78
19	0.99985	3.845	12	0.2938	78
20	0.9999	4.027	12	0.3117	78
21	0.99993	4.187	12	0.3267	92
22	0.99996	4.438	12	0.3486	92
23	0.99998	4.750	12	0.3734	106
24	0.999985	4.880	12	0.3829	107
25	0.99999	5.063	12	0.3954	107
26	0.999993	5.224	12	0.4056	107
27	0.999996	5.477	12	0.4202	121
28	0.999998	5.790	12	0.4359	122

Table 8.2   System Fluctuation Characteristic Parameters Corresponding to Different (α_1, β_1) Combinations When *MTBF* = 28.5179

Experiment	Parameter				
	α_1	β_1	T_0	M	T
1	0.92	0.779	6	0.063	53
2	0.93	0.814	6	0.0506	44
3	0.94	0.855	7	0.0381	35
4	0.95	0.904	7	0.0254	24
5	0.96	0.964	8	0.0116	12
6	0.97	1.042	590	0	0
7	0.98	1.153	165	0	0

(Continued)

Table 8.2 *(Continued)* **System Fluctuation Characteristic Parameters Corresponding to Different (α_1, β_1) Combinations When *MTBF* = 28.5179**

Experiment	Parameter				
	α_1	β_1	T_0	M	T
8	0.99	1.345	67	0.0001	0
9	0.993	1.445	52	0.0003	0
10	0.996	1.603	42	0.0014	0
11	0.997	1.685	38	0.0025	0
12	0.999	2	33	0.0095	43
13	0.9993	2.102	33	0.0127	44
14	0.9995	2.2	32	0.016	44
15	0.9997	2.348	32	0.0217	44
16	0.99973	2.379	32	0.0229	44
17	0.99976	2.413	32	0.0243	44
18	0.99979	2.452	32	0.026	44
19	0.99983	2.514	32	0.0287	43
20	0.99985	2.615	32	0.0332	43
21	0.9999	2.669	32	0.0357	73
22	0.99996	2.937	32	0.0489	78
23	0.999985	3.226	32	0.0642	78
24	0.999993	3.451	32	0.0768	110
25	0.999998	3.821	32	0.0981	112

Table 8.3 **System Fluctuation Characteristic Parameters Corresponding to Different (α_1, β_1) Combinations When *MTBF* = 40.1283**

Experiment	Parameter				
	α_1	β_1	T_0	M	T
1	0.92	0.716	6	0.0818	86
2	0.93	0.748	6	0.0693	73
3	0.94	0.785	6	0.0557	61

(Continued)

Table 8.3 *(Continued)* **System Fluctuation Characteristic Parameters Corresponding to Different (α_1, β_1) Combinations When *MTBF* = 40.1283**

Experiment	Parameter				
	α_1	β_1	T_0	M	T
4	0.95	0.829	7	0.0414	48
5	0.96	0.883	7	0.027	34
6	0.97	0.954	8	0.0109	13
7	0.98	1.054	N	0	0
8	0.99	1.227	153	0	0
9	0.993	1.318	107	0	0
10	0.996	1.46	76	0.0002	0
11	0.997	1.534	67	0.0005	0
12	0.999	1.818	52	0.0028	0
13	0.9993	1.911	50	0.0042	0
14	0.9995	2	48	0.0057	54
15	0.9997	2.134	47	0.0083	58
16	0.99973	2.161	47	0.0089	58
17	0.99976	2.192	46	0.0096	59
18	0.99979	2.227	46	0.0104	59
19	0.99983	2.283	46	0.0117	59
20	0.99985	2.316	46	0.0126	59
21	0.9999	2.423	45	0.0154	59
22	0.99996	2.666	44	0.0226	59
23	0.999985	2.927	44	0.0314	101
24	0.999993	3.131	44	0.0388	104
25	0.999998	3.466	44	0.0519	105

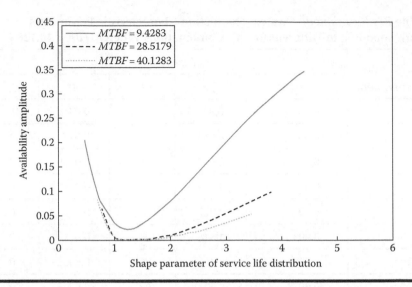

Figure 8.1 **Availability amplitude comparison when *MTBF* = 9.4283, 28.5179, 40.1283.**

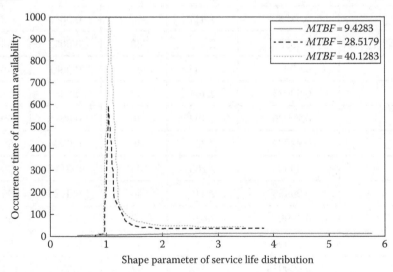

Figure 8.2 **Comparison for occurrence time of minimum availability when *MTBF* = 9.4283, 28.5179, 40.1283.**

As seen in Figure 8.1, when $\alpha_2 = 0.995$ and $\beta_2 = 3$, as β_1 increases, the availability amplitude *M* first increases and then decreases, with its extreme value occurring near $\beta_1 = 1$; $\beta_1 = 1$ means the system service life follows the geometric distribution and the system failure rate is a constant. When *MTBF* = 40.1283 and *MTBF* = 28.5179, there are α_1 and β_1 combinations that can eliminate the system availability fluctuation.

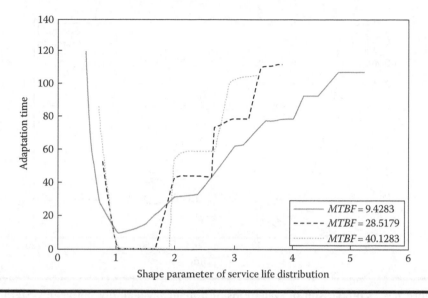

Figure 8.3 Adaptation time comparison when *MTBF* = 9.4283, 28.5179, 40.1283.

When the shape parameter β_1 is near 1, the system availability amplitude M has the minimum 0, which means the system instantaneous availability always is maintained above the steady-state availability value, and this kind of fluctuation becomes greater as it decreases; when $MTBF = 9.4283$, any combinations of α_1 and β_1 will impossibly eliminate the fluctuation of system instantaneous availability.

As seen from Table 8.3, when $\alpha_1 = 0.98$ and $\beta_1 = 1.054$, the system instantaneous availability monotonically decreases to steady-state availability, which means the minimum availability occurs at $T_0 = N$, when the system is matched. As the greatest time can be regarded as the infinite time in the actual engineering, then T_0 corresponding to $T_0 = N$ is selected as the upper boundary of 1000 of the research time period herein, as shown in Figure 8.2. With $MTBF$ fixed to any value, the occurrence time of minimum availability does not vary much after β_1 is set to 2, as is depicted in Figure 8.2; when $MTBF = 40.1283$ and $MTBF = 28.5179$, the occurrence time of minimum availability T_0 has the maximum value and sometimes may even go toward infinity. As $MTBF$ reaches 40.1283 or above, there are combinations α_1 and β_1 that can make the minimum availability occurrence time of the system T_0 reach infinity, and the system is matched at this time. When $MTBF = 9.4283$ or below, the occurrence time of minimum availability T_0 does not vary much.

As seen from Figure 8.2, when $MTBF = 9.4283$, no combinations of α_1 and β_1 in the system can make the system adaptation time zero, and the system cannot be matched by any parameters. When $MTBF = 28.5179$ and $MTBF = 40.1283$, there are combinations of α_1 and β_1 that can make the system to be matched. It can also be noted that, when $MTBF$ increases, the parameter selection range for system

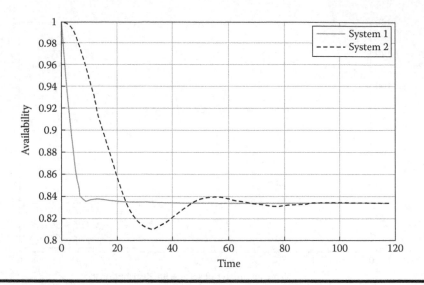

Figure 8.4 Comparison of system instantaneous availability between system 1 and system 2 with same *MTBF*.

matching also gradually enlarges. As seen from Figure 8.3, with *MTBF* fixed to any value, as T increases, the adaptation time first monotonically decreases and then reaches its minimum when β_1 is near 1. Then, it gradually increases.

To illustrate this, the instantaneous availabilities of two systems are compared when *MTBF* = 28.5179, α_2 = 0.995, and β_2 = 3, where α_1 = 0.97 and β_1 = 1.042 in system 1, and α_1 = 0.9997 and β_1 = 2.348 in system 2. The result is shown in Figure 8.4. Both systems have the same *MTBF* and the same distribution for the repair time to follow. But, as seen in the figure, different combinations of α_1 an β_1 can greatly change the instantaneous availability of systems, especially the fluctuation of instantaneous availability at the preliminary operation period. Table 8.2, Figure 8.1, and Figure 8.3 present related parameters. Therefore, it is believed that the availability amplitude, occurrence time of minimum availability, and the adaptation time can accurately describe the system fluctuation.

8.3.2 Effects of Scale and Shape Parameters in Repair Time Distribution on Instantaneous Availability Fluctuation

Shaping parameter β_1 and scale parameter α_1 in fixed service life distribution, and mean time to repair *MTTR*, suppose α_1 = 0.99995 and β_1 = 3, then *MTBF* = 24.739, with *MTTR* fixed to a value. Divide the simulation into three sets as *MTTR* varies:

1. Suppose *MTTR* = 4.6986, then the inherent availability A = 0.8404. Match α_1 and β_1, then obtain the fluctuation characteristic parameter values for the system, as shown in Table 8.4.

Table 8.4 System Fluctuation Characteristic Parameters Corresponding to Different (α_1, β_1) Combinations When *MTTR* = 4.6986

Experiment	Parameter				
	α_1	β_1	T_0	M	T
1	0.5	0.504	31	0.0077	34
2	0.6	0.629	30	0.0223	38
3	0.7	0.795	30	0.0337	39
4	0.8	1.037	29	0.0418	67
5	0.9	1.467	28	0.0467	67
6	0.91	1.533	28	0.0471	67
7	0.92	1.608	28	0.0474	67
8	0.93	1.693	28	0.0477	67
9	0.94	1.792	28	0.048	67
10	0.95	1.91	28	0.0482	67
11	0.96	2.055	28	0.0484	67
12	0.97	2.243	28	0.0486	67
13	0.98	2.51	27	0.0488	67
14	0.99	2.97	27	0.0492	67
15	0.993	3.209	27	0.0493	67
16	0.996	3.585	27	0.0494	67
17	0.999	4.525	27	0.0496	67
18	0.9993	4.769	27	0.0496	67
19	0.9996	5.151	27	0.0497	67
20	0.9998	5.627	27	0.0497	67

2. Suppose *MTTR* = 5.7178, then the inherent availability *A* = 0.812. Match α_1 and β_1, then obtain the fluctuation characteristic parameter values for the system, as shown in Table 8.5.

3. Suppose *MTTR* = 13.0174, then the inherent availability *A* = 0.6552. Match α_1 and β_1, then obtain the fluctuation characteristic parameter values for the system, as shown in Table 8.6.

Table 8.5 System Fluctuation Characteristic Parameters Corresponding to Different (α_1, β_1) Combinations When *MTTR* = 5.7178

Experiment	Parameter				
	α_1	β_1	T_0	M	T
1	0.6	0.574	31	0.014	37
2	0.7	0.719	31	0.0311	39
3	0.8	0.929	30	0.0451	69
4	0.9	1.3	29	0.0555	70
5	0.91	1.357	29	0.0564	70
6	0.92	1.422	29	0.0571	70
7	0.93	1.495	29	0.0579	70
8	0.94	1.581	29	0.0585	70
9	0.95	1.682	28	0.0593	70
10	0.96	1.807	28	0.0601	70
11	0.97	1.969	28	0.0608	70
12	0.98	2.2	28	0.0615	69
13	0.99	2.598	28	0.0622	69
14	0.993	2.804	28	0.0624	69
15	0.996	3.13	28	0.0627	69
16	0.999	3.944	28	0.063	69
17	0.9993	4.155	28	0.0631	69
18	0.9996	4.487	28	0.0631	69
19	0.9998	4.899	28	0.0632	69

Table 8.6 System Fluctuation Characteristic Parameters Corresponding to Different (α_1, β_1) Combinations When *MTTR* = 13.0174

Experiment	Parameter				
	α_1	β_1	T_0	M	T
1	0.8	0.668	34	0.0105	38
2	0.9	0.907	33	0.0603	73

(Continued)

Table 8.6 *(Continued)* System Fluctuation Characteristic Parameters Corresponding to Different (α_1, β_1) Combinations When *MTTR* = 13.0174

Experiment	Parameter				
	α_1	β_1	T_0	M	T
3	0.91	0.944	33	0.0661	75
4	0.92	0.985	33	0.0722	76
5	0.93	1.032	33	0.0785	78
6	0.94	1.087	33	0.0853	79
7	0.95	1.152	33	0.0924	80
8	0.96	1.232	33	0.1002	82
9	0.97	1.336	32	0.109	83
10	0.98	1.484	32	0.1196	84
11	0.99	1.74	32	0.1332	84
12	0.993	1.873	32	0.1385	117
13	0.996	2.084	32	0.1451	120
14	0.999	2.61	31	0.1568	121
15	0.9993	2.747	31	0.1589	121
16	0.9996	2.962	31	0.1617	120
17	0.9998	3.23	31	0.1644	120
18	0.99983	3.292	31	0.1651	120
19	0.99986	3.368	31	0.1656	120
20	0.99989	3.461	31	0.1664	120

The data results in Tables 8.4 through 8.6 are graphed in Figures 8.5 through 8.7. The x-coordinate is the shape parameter β_2 of repair time distribution, and the y-coordinate is each kind of fluctuation parameter.

As seen in Figure 8.5, with *MTTR* fixed to any value, as β_2 increases, the system availability amplitude M gradually increases. When β_2 reaches a certain specified value, the availability amplitude M becomes steady, and the corresponding specified value also increases as *MTTR* increases. As seen in Figure 8.6, with *MTTR* fixed to any value, as β_2 increases, the occurrence time of minimum availability

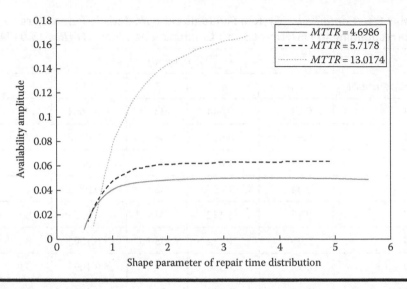

Figure 8.5   Availability amplitude comparison when *MTTR* = 4.6986, 5.7178, 13.0174.

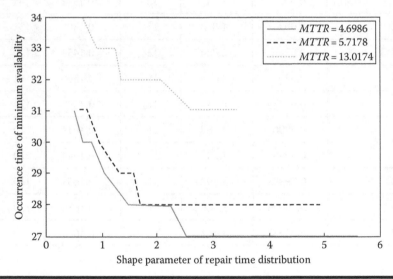

Figure 8.6   Comparison for occurrence time of minimum availability when *MTTR* = 4.6986, 5.7178, 13.0174.

T_0 gradually decreases and becomes steady at a certain specified value. The corresponding specified value increases as *MTTR* increases. As seen in Figure 8.7, with *MTTR* fixed to any value, as β_2 increases, the system adaptation T gradually decreases and becomes steady at a certain specified value. The corresponding specified value increases as *MTTR* increases.

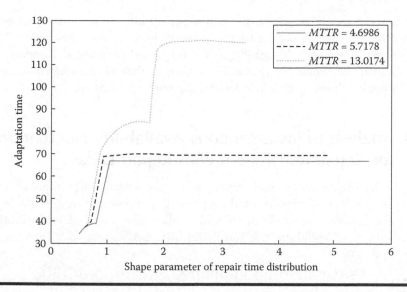

Figure 8.7   Adaptation time comparison when *MTTR* = 4.6986, 5.7178, 13.0174.

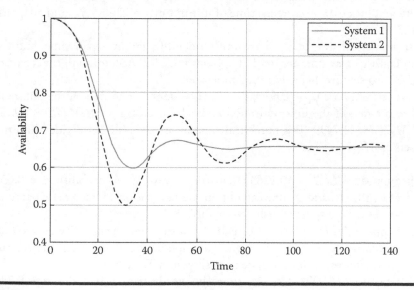

Figure 8.8   Comparison of system instantaneous availability between system 1 and system 2 with same *MTTR*.

To illustrate this, the instantaneous availabilities of two systems are compared when $MTTR = 13.0174$, $\alpha_1 = 0.99995$, and $\beta_1 = 3$, where $\alpha_2 = 0.9$ and $\beta_2 = 0.907$ in system 1, and $\alpha_2 = 0.9993$ and $\beta_2 = 2.747$ in system 2. The result is shown in Figure 8.8. Both systems have the same *MTTR* and completely the same distribution for the unit service life to follow. But as seen in the figure, different combinations of

α_2 and β_2 can greatly change the instantaneous availability of systems, especially the fluctuation of instantaneous availability at the preliminary operation period. Refer to Table 8.6, Figure 8.5 through Figure 8.7 for related parameters. Therefore, it is believed that the availability amplitude, occurrence time of minimum availability, and adaptation time can accurately describe the system fluctuation.

8.4 Analysis of Instantaneous Availability Fluctuation for Repairable System with Repair Delay

For the repairable system with repair delay, the system instantaneous availability $A(k)$ follows Formulas 5.8 through 5.10. Suppose the service life, repair time, and logistic delay time of the units follow the probability distributions Weibull(α_1, β_1), Weibull(α_2, β_2), and Weibull(α_3, β_3), respectively. Then,

$$I = I\big(A(.)\big) = I\big(\lambda(.), \mu(.), \rho(.)\big) = I\big(\alpha_1, \beta_1, \alpha_2, \beta_2, \alpha_3, \beta_3\big),$$

where I is a six-variable function, where the three variables selected here are availability amplitude M, occurrence time of minimum availability T_0, and adaptation time T.

In numerical simulation, fixed unit service life, failure repair time, and fixed mean logistic delay time are α_1, β_1, α_2, β_2 and $MLDT$. According to the repair rate $MLDT$, two sets are divided in the simulation:

1st set: Suppose $\alpha_1 = 0.9999$, $\beta_1 = 2$, $\alpha_2 = 0.9995$, and $\beta_2 = 3$; the failure rate and repair rate of the unit both monotonically decrease, with $MTBF = 89.1205$ and $MTTR = 11.7499$. This numerical simulation set is further divided into three subsets as $MLDT$ varies:

1. Suppose $MLDT = 40.1283$, then the steady-state availability $A = 0.6321$. Matching α_3 and β_3, one can obtain the fluctuation characteristic parameter values for the system, as shown in Table 8.7.
2. Suppose $MLDT = 28.5179$, then the steady-state availability $A = 0.6888$. Matching α_3 and β_3, then one can obtain the fluctuation characteristic parameter values for the system, as shown in Table 8.8.
3. Suppose $MLDT = 13.0174$, then the steady-state availability $A = 0.7825$. Matching each α_3 and β_3, then one can obtain the fluctuation characteristic parameter values for the system, as shown in Table 8.9.

The data results in Tables 8.7 through 8.9 are graphed in Figures 8.9 through 8.11. The x-coordinate is the shape parameter β_3 of logistic repair time distribution, and the y-coordinate is each kind of fluctuation parameter I.

As seen in Figure 8.9, with $MLDT$ fixed to any value, the availability amplitude M monotonically increases as β_2 increases with slow growth. After β_2 reaches a

Table 8.7 System Fluctuation Characteristic Parameters Corresponding to Different (α_1, β_1) Combinations When *MLDT* = 40.1283

Experiment	Parameter				
	α_1	β_1	T_0	M	T
1	0.9	0.663	N	0	0
2	0.93	0.748	137	0.0082	157
3	0.95	0.8290	133	0.0169	172
4	0.96	0.8830	131	0.0217	176
5	0.97	0.954	128	0.0272	178
6	0.98	1.0540	125	0.0336	178
7	0.99	1.2270	121	0.0419	175
8	0.993	1.3180	120	0.0450	173
9	0.995	1.4030	118	0.0477	171
10	0.997	1.5340	116	0.0509	169
11	0.999	1.8180	113	0.0557	165
12	0.9993	1.9110	113	0.0568	164
13	0.9995	2.0000	112	0.0576	163
14	0.9997	2.1340	111	0.0587	162
15	0.9998	2.2400	111	0.0595	161

Table 8.8 System Fluctuation Characteristic Parameters Corresponding to Different (α_1, β_1) Combinations When *MLDT* = 28.5179

Experiment	Parameter				
	α_1	β_1	T_0	M	T
1	0.9	0.7200	130	0.0099	156
2	0.93	0.8140	125	0.0168	165
3	0.95	0.9040	122	0.0218	167
4	0.96	0.9640	120	0.0244	167
5	0.97	1.0420	118	0.0272	166

(Continued)

Table 8.8 *(Continued)* **System Fluctuation Characteristic Parameters Corresponding to Different (α_1, β_1) Combinations When** *MLDT* **= 28.5179**

Experiment	Parameter				
	α_1	β_1	T_0	M	T
6	0.98	1.1530	116	0.0302	164
7	0.99	1.3450	112	0.0335	161
8	0.993	1.4450	111	0.0347	159
9	0.995	1.5400	110	0.0355	158
10	0.997	1.6850	109	0.0364	157
11	0.999	2.0000	107	0.0375	154
12	0.9993	2.1020	106	0.0377	154
13	0.9995	2.2000	106	0.0379	153
14	0.9997	2.3480	105	0.0380	153
15	0.9998	2.4660	105	0.0382	153

Table 8.9 System Fluctuation Characteristic Parameters Corresponding to Different (α_1, β_1) Combinations When *MLDT* **= 13.0174**

Experiment	Parameter				
	α_1	β_1	T_0	M	T
1	0.65	0.4800	126	0.0005	0
2	0.70	0.5320	123	0.0044	0
3	0.75	0.5930	120	0.0076	141
4	0.80	0.6680	117	0.0104	147
5	0.85	0.7660	114	0.0126	149
6	0.90	0.9070	111	0.0142	148
7	0.93	1.0320	109	0.0149	147
8	0.95	1.1520	108	0.0152	146
9	0.96	1.2320	107	0.0153	145
10	0.97	1.3360	107	0.0153	145

(Continued)

Table 8.9 *(Continued)* **System Fluctuation Characteristic Parameters Corresponding to Different (α_1, β_1) Combinations When *MLDT* = 13.0174**

Experiment	Parameter				
	α_1	β_1	T_0	M	T
11	0.98	1.4840	106	0.0154	144
12	0.99	1.7400	105	0.0154	144
13	0.993	1.8730	105	0.0154	143
14	0.995	2.0000	105	0.0153	143
15	0.997	2.1920	105	0.0153	143

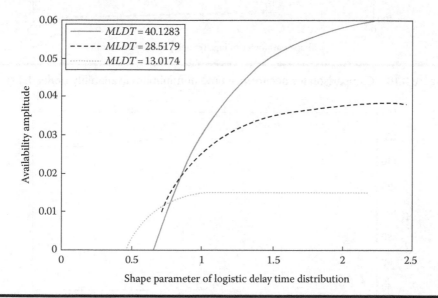

Figure 8.9   Availability amplitude comparison under 1st set.

certain critical value, the availability becomes steady at a certain specified value. The corresponding critical value increase as *MLDT* increases, and so does the specified value. As *MLDT* increases, the change rate of availability amplitude corresponding to β_2 also increases.

As seen from Figure 8.10, with *MLDT* fixed to any value, the occurrence time of system minimum availability does not vary much with β_3 change. But when *MLDT* = 40.1283 except β_3 < 0.748, the occurrence time of system minimum availability increases as *MLDT* increases. When *MLDT* = 40.1283 and β_3 = 0.663, the occurrence time of system minimum availability is at infinity, and the system is matched at this time.

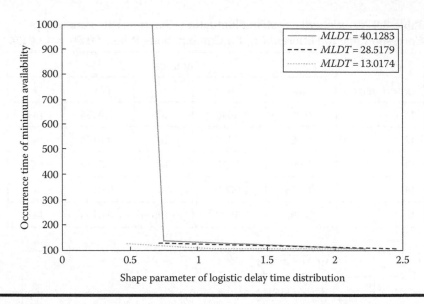

Figure 8.10   Comparison for occurrence time of minimum availability under 1st set.

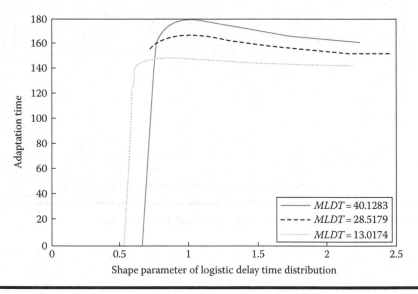

Figure 8.11   Adaptation time comparison under 1st set.

As seen from Figure 8.11, with *MLDT* fixed to any value, the system adaptation time is 0 in cases of small β_3, and now the system is matched. But as β_3 increases, the system instantaneous availability fluctuates more strongly, and the system adaptation time reaches the extreme value when β_3 is at certain critical value. As *MLDT* increases, the corresponding critical value increases, and so does the extreme value.

As β_3 further increases to another critical value, the system adaptation time gradually becomes steady at a certain specified value. As *MLDT* increases, the corresponding critical value increases, and so does the specified value.

2nd set: Suppose $\alpha_1 = 0.99998$, $\beta_1 = 2$, $\alpha_2 = 0.95$, and $\beta_2 = 0.85$; the failure rate of the unit monotonically increases, and its repair rate monotonically decreases, with *MTBF* = 198.6654 and *MTTR* = 36.3307. This numerical simulation set is further divided into three subsets as *MLDT* varies:

1. Suppose *MLDT* = 40.1283, then the steady-state availability $A = 0.7221$. Matching α_3 and β_3, one can then obtain the fluctuation characteristic parameter values for the system, as shown in Table 8.10.
2. Suppose *MLDT* = 28.5179, then the steady-state availability $A = 0.7539$. Matching α_3 and β_3, one can then obtain the fluctuation characteristic parameter values for the system, as shown in Table 8.11.

Table 8.10 System Fluctuation Characteristic Parameters Corresponding to Different (α_1, β_1) Combinations When *MLDT* = 40.1283

Experiment	Parameter				
	α_1	β_1	T_0	M	T
1	0.8	0.5010	0	304	0.0010
2	0.85	0.5680	335	295	0.0072
3	0.9	0.6630	365	286	0.0132
4	0.93	0.7480	371	279	0.0167
5	0.95	0.8290	371	274	0.0188
6	0.96	0.8830	370	271	0.0199
7	0.97	0.9540	368	268	0.0208
8	0.98	1.0540	365	264	0.0217
9	0.99	1.2270	361	260	0.0225
10	0.993	1.3180	359	259	0.0226
11	0.995	1.4030	357	257	0.0228
12	0.997	1.5340	356	256	0.0228
13	0.999	1.8180	353	254	0.0228
14	0.9993	1.9110	352	254	0.0228
15	0.9995	2.0000	352	253	0.0227
16	0.9997	2.1340	351	253	0.0227

Table 8.11 System Fluctuation Characteristic Parameters Corresponding to Different (α_1, β_1) Combinations When *MLDT* = 28.5179

	Parameter				
Experiment	α_1	β_1	T_0	M	T
1	0.7	0.4370	0	299	0.0006
2	0.75	0.4830	0	293	0.0046
3	0.8	0.5400	339	287	0.0083
4	0.85	0.6140	355	280	0.0115
5	0.9	0.7200	359	273	0.0142
6	0.93	0.8140	358	269	0.0155
7	0.95	0.9040	356	266	0.0161
8	0.96	0.9640	354	264	0.0163
9	0.97	1.0420	353	262	0.0165
10	0.98	1.1530	351	261	0.0166
11	0.99	1.3450	349	259	0.0166
12	0.993	1.4450	348	258	0.0166
13	0.995	1.5400	347	258	0.0165
14	0.997	1.6850	346	257	0.0165
15	0.999	2.0000	345	256	0.0164
16	0.9993	2.1020	345	256	0.0164
17	0.9995	2.2000	345	256	0.0164
18	0.9997	2.3480	344	256	0.0164
19	0.9998	2.4660	344	256	0.0164

3. Suppose *MLDT* = 13.0174, then the steady-state availability A = 0.8010. Matching α_3 and β_3, one can then obtain the fluctuation characteristic parameter values for the system, as shown in Table 8.12.

The data results in Tables 8.10 through 8.12 are graphed in Figures 8.12 through 8.14. The x-coordinate is the shape parameter β_3 of logistic repair time distribution, and the y-coordinate is each kind of fluctuation parameter I.

Table 8.12   System Fluctuation Characteristic Parameters Corresponding to Different (α_1, β_1) Combinations When *MLDT* = 13.0174

Experiment	Parameter				
	α_1	β_1	T_0	M	T
1	0.60	0.4360	326	280	0.0072
2	0.65	0.4800	334	277	0.0083
3	0.70	0.5320	337	274	0.0091
4	0.75	0.5930	338	272	0.0096
5	0.80	0.6680	338	270	0.0098
6	0.85	0.7660	336	268	0.0100
7	0.90	0.9070	335	267	0.0100
8	0.93	1.0320	334	266	0.0099
9	0.95	1.1520	334	266	0.0099
10	0.96	1.2320	333	266	0.0099
11	0.97	1.3360	333	265	0.0099
12	0.98	1.4840	333	265	0.0099
13	0.99	1.7400	333	265	0.0098
14	0.993	1.8730	332	265	0.0098
15	0.995	2.0000	332	265	0.0098
16	0.997	2.1920	332	265	0.0098

As seen in Figure 8.12, with *MLDT* fixed to any value, the availability amplitude M monotonically increases as β_2 increases with slow growth. After β_2 reaches a certain critical value, the availability becomes steady at a certain specified value. The corresponding critical value increases as *MLDT* increases, and so does the specified value. As *MLDT* increases, the change rate of availability amplitude corresponding to β_2 also increases. This is similar to the result shown in Figure 8.9.

As seen from Figure 8.13, with *MLDT* fixed, as β_3 increases, α_3 increases. Meanwhile, the occurrence time of minimum availability decreases, and this decrease amplitude also decreases. When β_3 increases to a certain critical value, the occurrence time of minimum availability decreases to a certain specified value and remains unchanged. As *MLDT* increases, the critical value also increases, but the corresponding specified value decreases.

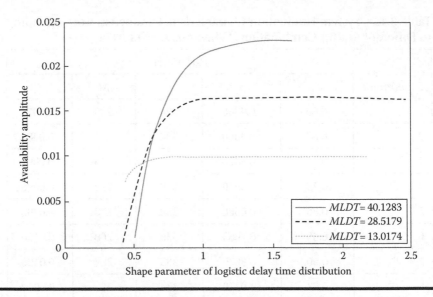

Figure 8.12   Availability amplitude comparison under 2nd set.

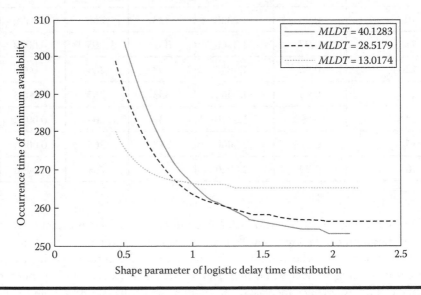

Figure 8.13   Comparison for occurrence time of minimum availability under 2nd set.

As seen from Figure 8.14, with *MLDT* fixed, the system adaptation time is 0 in cases of small β_3, and now the system is matched. But as β_3 increases, the system instantaneous availability fluctuates more greatly, and the system adaptation time reaches the extreme value when β_3 is at a certain critical value. As *MLDT* increases, the corresponding critical value increases, and so does the extreme value.

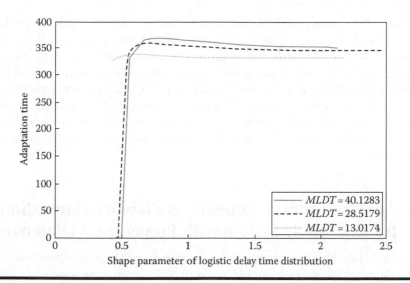

Figure 8.14 **Adaptation time comparison under 2nd set.**

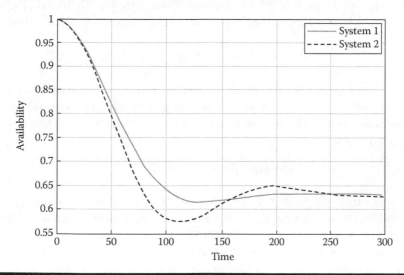

Figure 8.15 **Comparison of system instantaneous availability between system 1 and system 2 with same *MLDT*.**

As β_3 further increases to another critical value, the system adaptation time gradually becomes steady at a certain specified value. As *MLDT* increases, the corresponding critical value increases, and so does the specified value.

To illustrate this, a subset of simulation analysis is selected from the 1st set: $\alpha_1 = 0.9999$, $\beta_1 = 2$, $\alpha_2 = 0.9995$, $\beta_2 = 3$, and *MLDT* = 40.1283. Comparison results for the instantaneous availability of two systems are shown in Figure 8.15, where $\alpha_3 = 0.95$ and $\beta_3 = 0.8290$ in system 1, and $\alpha_3 = 0.9995$ and $\beta_3 = 2.0000$ in system 2. Both systems have the same *MLDT* and completely the same distribution for the service life and repair time to follow. But, as seen in Figure 8.15, different combinations of α_2 and β_2 can greatly change the instantaneous availability of systems, especially the fluctuation of instantaneous availability at the preliminary operation period. Refer to Table 8.7, Figures 8.9 through 8.11 for related parameters. Therefore, it is believed that the availability amplitude, occurrence time of minimum availability, and the adaptation time can accuratey describe the system fluctuation.

8.5 Analysis of Instantaneous Availability Fluctuation for Repairable System with Preventive Maintenance

For a repairable system with preventive maintenance, the system instantaneous availability $A(k)$ meets Formulas 5.15 through 5.17. Suppose the service life, repair time for repairable maintenance, and repair time for preventive maintenance of the units follow the probability distributions Weibull(α_1, β_1), Weibull(α_2, β_2), and Weibull(α_3, β_3), respectively, and the preventive maintenance cycle is N. Then,

$$I = I\left(A(.)\right) = I\left(\lambda(.),\mu(.)\right) = I\left(\alpha_1,\beta_1,\alpha_2,\beta_2\,\alpha_3,\beta_3,N\right)$$

where I is a seven-variable function, and the variable selected here is the mean availability or interval availability within a limited period. The time interval selected here for study is $[0,150]$, and the research index is the mean availability within $[0,150]$, written as

$$\bar{A} = \bar{A}\left(\alpha_1,\beta_1,\alpha_2,\beta_2,\alpha_3,\beta_3,N\right).$$

Suppose $\alpha_2 = 0.999$, $\beta_2 = 2$, scale parameter of fixed service life distribution $\alpha_1 = 0.99998$, shape parameter $\beta_1 = 3$, and mean preventive maintenance time *MPMT*. Then, *MTBF* $= 33.3975$ and *MTTR* $= 28.5179$. To study the effects of repair time distribution for preventive maintenance and for the preventive maintenance cycle on the fluctuation, two groups are divided in numerical simulation as follows.

8.5.1 Effects of Repair Time Distribution for Preventive Maintenance on System Availability Fluctuation

Suppose $\alpha_1 = 0.99998$, $\beta_1 = 3$, and $N = 30$, with *MPMT* fixed to a value. Divide the simulation into three sets as *MPMT* varies:

1. Suppose *MPMT* $= 13.0174$, matching α_3 and β_3, and then one can obtain the system fluctuation characteristic parameters as shown in Table 8.13.

Table 8.13 System Fluctuation Characteristic Parameters Corresponding to Different (α_1, β_1) Combinations When MPMT = 13.0174

Experiment	α_3	β_3	$\bar{A}$
1	0.65	0.4800	0.6362
2	0.7	0.5320	0.6329
3	0.75	0.5930	0.6297
4	0.8	0.6680	0.6268
5	0.85	0.7660	0.6245
6	0.9	0.9070	0.6228
7	0.93	1.0320	0.6217
8	0.95	1.1520	0.6211
9	0.96	1.2320	0.6208
10	0.97	1.3360	0.6205
11	0.98	1.4840	0.6202
12	0.99	1.7400	0.6200
13	0.993	1.8730	0.6200
14	0.995	2.0000	0.6201
15	0.997	2.1920	0.6200
16	0.999	2.6100	0.6202
17	0.9993	2.7470	0.6203
18	0.9995	2.8760	0.6204
19	0.9997	3.0730	0.6205
20	0.9998	3.2300	0.6206

2. Suppose $MPMT$ = 9.3400, matching α_3 and β_3, one can then obtain the system fluctuation characteristic parameters as shown in Table 8.14.
3. Suppose $MPMT$ = 4.413, matching α_3 and β_3, one can then obtain the system fluctuation characteristic parameters as shown in Table 8.15.

The data results in Tables 8.13 through 8.15 are graphed in Figure 8.16. The x-coordinate is the shape parameter β_3 under the preventive maintenance time

Table 8.14 System Fluctuation Characteristic Parameters Corresponding to Different (α_1, β_1) Combinations When MPMT = 9.3400

Experiment	α_3	β_3	$\bar{A}$
1	0.65	0.5310	0.6531
2	0.7	0.5910	0.6514
3	0.75	0.6610	0.6497
4	0.8	0.7490	0.6486
5	0.85	0.8640	0.6477
6	0.9	1.0270	0.6469
7	0.93	1.1740	0.6466
8	0.95	1.3130	0.6464
9	0.96	1.4070	0.6464
10	0.97	1.5280	0.6463
11	0.98	1.7010	0.6463
12	0.99	2.000	0.6464
13	0.993	2.1550	0.6464
14	0.995	2.3020	0.6464
15	0.997	2.5260	0.6464
16	0.999	3.0130	0.6465
17	0.9993	3.1720	0.6465
18	0.9995	3.3220	0.6465
19	0.9997	3.5510	0.6466
20	0.9998	3.7330	0.6466

Table 8.15 System Fluctuation Characteristic Parameters Corresponding to Different (α_1, β_1) Combinations When MPMT = 4.413

Experiment	α_3	β_3	$\bar{A}$
1	0.65	0.7300	0.6868
2	0.7	0.8240	0.6865
3	0.75	0.9380	0.6863

(Continued)

Table 8.15 *(Continued)*   **System Fluctuation Characteristic Parameters Corresponding to Different (α_1, β_1) Combinations When *MPMT* = 4.413**

Experiment	α_3	β_3	$\overline{A}$
4	0.8	1.0790	0.6861
5	0.85	1.2650	0.6859
6	0.9	1.5320	0.6858
7	0.93	1.7710	0.6857
8	0.95	2.0000	0.6857
9	0.96	2.1530	0.6857
10	0.97	2.3510	0.6857
11	0.98	2.6320	0.6856
12	0.99	3.1170	0.6856
13	0.993	3.3680	0.6856
14	0.995	3.6060	0.6856
15	0.997	3.9690	0.6856

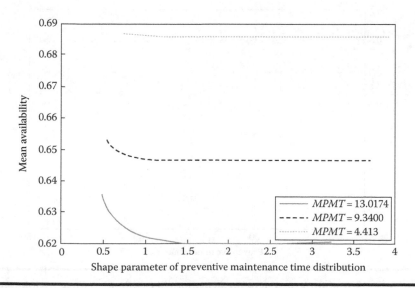

Figure 8.16   Mean availability comparison.

distribution, and the y-coordinate is the system mean availability. As can easily be seen in Figure 8.16, when the preventive maintenance cycle remains unchanged, *MPMT* increases, and then the system mean availability decreases. Also, with *MPMT* fixed, when $\beta_3 \geq 1$, the system mean availability is not greatly affected by the parameter under preventive maintenance time distribution; but when β_3 is to the left of 1, this effect may grow bigger; as β_3 and α_3 decrease, the system mean availability increases accordingly. This variation will become more obvious as *MPMT* increases.

8.5.2 Effects of Preventive Maintenance Cycle on System Availability Fluctuation

Suppose $\alpha_3 = 0.995$ and $\beta_3 = 3$. Two sets are also divided in the simulation as a result of a different failure rate monotonicity for the systems:

1. System 1: Suppose $\alpha_1 = 0.99998$ and $\beta_1 = 3$. Now the system failure rate increases monotonically. The variation of system mean availability with preventive maintenance cycle N is shown in Figure 8.17.
2. System 2: Suppose $\alpha_1 = 0.998$ and $\beta_1 = 3$. Now the system failure rate decreases monotonically. The variation of system mean availability with preventive maintenance cycle N is shown in Figure 8.18.

As seen in Figure 8.17, the system mean availability increases as the preventive maintenance cycle N increases, and then decreases with a maximum value when

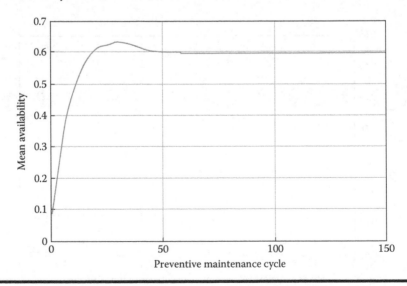

Figure 8.17 Variation 1 of system mean availability with preventive maintenance cycle.

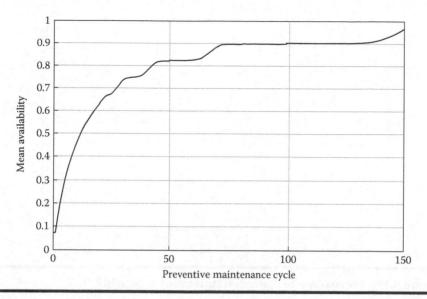

Figure 8.18   Variation 2 of system mean availability with preventive maintenance cycle.

the preventive maintenance cycle is 29. At this moment, from the mean availability viewpoint, 29 can be considered as the optimal preventive maintenance cycle. When the preventive maintenance cycle exceeds 50, it has almost no effect on the system mean availability.

As seen in Figure 8.18, the system mean availability increases as the preventive maintenance cycle N increases. When the preventive maintenance cycle is large enough and, especially, when it exceeds the research time interval, it can be interpreted as there being no preventive maintenance in the system. This system is degraded to a normal single-unit repairable system.

It can be noted from the comparison between Figure 8.17 and Figure 8.18 that, from the mean availability view, system 1 has the optimal preventive maintenance cycle. But system 2 either does not have optimal preventive maintenance, or it can be interpreted as no preventive maintenance applied in this system. This is because the preventive maintenance can turn back the service age. For the system with an increasing failure rate, preventive maintenance reduces the system failure rate and then improves the system's mean availability. But for the system with a decreasing failure rate, preventive maintenance means increasing the system failure rate. For this reason, preventive maintenance for the system with increasing failure rate is one way of improving the system mean availability.

To illustrate this, we suppose $\alpha_1 = 0.99998$, $\beta_1 = 3$, $\alpha_2 = 0.999$, $\beta_2 = 2$, $\alpha_3 = 0.995$, and $\beta_3 = 3$, and we also compare the instantaneous availability of two systems with different preventive maintenance cycles—where $N = 29$ in system 1, and $N = 55$ in system 2.

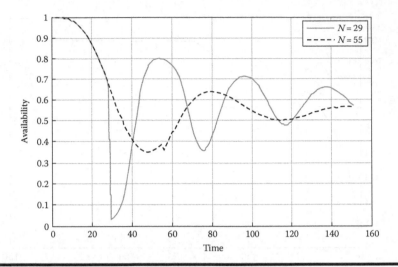

Figure 8.19 **Comparison of system instantaneous availability.**

The result is shown in Figure 8.19. The service life, repair time for repairable maintenance, and repair time for preventive maintenance of the two systems, all follow the same distribution, but the different preventive maintenance cycles greatly affect the accumulation of mean availability fluctuation, as shown in Figure 8.17. As shown in Figure 8.19, the preventive maintenance makes the instantaneous availability fluctuate more heavily. This fluctuation becomes more obvious as the preventive maintenance cycle decreases, which is caused by frequent down-time without failure. Although the system instantaneous availability fluctuates more heavily as the preventive maintenance cycle decreases, the service age advance can reduce the system failure rate and thus improve the system mean availability. At this moment, it is obviously appropriate to describe the availability fluctuation with the availability fluctuation amplitude, occurrence time of minimum availability, and with adaptation time. Therefore, in this case, the mean availability is used as the parameter for research.

8.6 Summary

This chapter simulates the instantaneous availability models for discrete-time single-unit repairable systems under truncated and discrete Weibull distribution to study the characteristics of instantaneous availability fluctuation. Conclusions are as follows:

1. It is not enough to evaluate the system performance by only considering *MTBF*, *MTTR*, *MLDT*, and *MPMT*. Because these parameters are fixed as constrained, the different parameter combinations under related time

probability distribution still greatly influence the fluctuation of system instantaneous availability. Therefore, it is reasonable to increase other parameters to study the fluctuation of system instantaneous availability.

2. Under the assumption of truncated and discrete Weibull distribution, when no down-time events without failure occur (such as preventive maintenance), the availability amplitude, occurrence time of minimum availability, and adaptation time can best describe the fluctuation characteristics of system instantaneous availability. When a preventive maintenance down-time event occurs, the mean availability can be used to describe the accumulation of instantaneous availability fluctuation. Therefore, the fluctuation parameters are selected depending on the real situations.

Some rules have been obtained from the research on availability fluctuation parameters for various classic repairable systems. They can help us in the design, analysis, and management of real engineering and have important significance for the system efficiency analysis and system availability design.

Chapter 9

Optimal Design of Instantaneous Availability Fluctuation under Truncated and Discrete Weibull Distribution

9.1 Overview

The fluctuation of equipment availability is affected by its own reliability, maintainability, supportability (RMS) characteristics, which are in turn affected by multiple factors occurring in the equipment's real use, maintenance, and support. To facilitate our study, this chapter assumes that all the time variables following the truncated and discrete Weibull distribution, and that the influencing factors for the system failure rate, repair rate, and logistic delay rate are shape parameter and scale parameter under truncated and discrete Weibull distribution. With those assumptions, the optimal control based on the fluctuation of system instantaneous availability turns to a multivariable optimization problem.

Chapter 8 covered simulation analysis of instantaneous availability fluctuation parameters under truncated and discrete Weibull distribution and outlined some rules. Based on these rules, the particle swarm optimization is used to optimally design the common parameters for instantaneous availability fluctuation in three phases—demonstration, development, and operation.

9.2 Algorithm Design

Generally, there are two methods used to solve the optimization: the deterministic method and the random method. For example, various conventionally analytic deterministic optimization algorithms have a quick convergence speed and high calculation accuracy, but the conditions of their objective functions are limited and the initial values are sensitive, which easily causes a local minimum [102]. In recent years, the random optimization algorithms, such as genetic algorithm, simulated annealing, and particle swarm optimization, have been widely applied in scientific research and engineering technology.

Particle Swarm Optimization (PSO) was first proposed by Dr. James Kennedy in social psychology and by Dr. Russell Eberhart in 1995 [103]. PSO is a kind of evolutionary computation technology—a swarm intelligence algorithm. Recent research and practices show that PSO has a quick convergence speed, a high solution quality, and robustness in multidimensional space function optimization and in dynamic target optimization; PSO has been widely used in engineering practices [101–103]. Therefore, this chapter uses PSO to solve the optimization problem.

In PSO, every particle is a solution in n-dimensional space. Suppose the position and velocity vectors of particle i are $X_i = (x_1, x_2, ..., x_n)$ and $V_i = (v_1, v_2, ..., v_n)$. The optimal position that this particle passes by is PB_i, and the optimal position that the swarm passes by is GB. Then the position and velocity updated equation of particle i is shown as follows:

$$\begin{cases} V_i^{k+1} = \omega^* V_i^k + c_1 * rand(.) * \left(PB_i^k - X_i^k \right) + c_2 * rand(.) * \left(GB^k - X_i^k \right) \\ X_i^{k+1} = X_i^k + V_i^{k+1}, \end{cases} \tag{9.1}$$

where V_i^k is the updated velocity of particle i at kth update, X_i^k is the position of particle i before kth update, ω is the inertia weight, and c_1 and c_2 are the acceleration constants. From the social psychology view, ω means the particle is dependent on its own current information, c_1 means the particle is dependent on its own experience, c_2 means the particle is dependent on community information, and $rand(.)$ represents the random numbers from 0 to 1.

The algorithm process flow is shown in Figure 9.1. For the models in this chapter, the algorithm process can be simplified with the particle fitness properly defined, as shown in Figure 9.2.

The algorithm steps are

Step 1: Initialize a N-sized particle swarm and set initial position and velocity.
Step 2: Calculate the fitness of every particle.

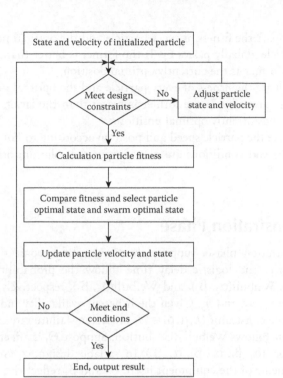

Figure 9.1   Algorithm process flow 1.

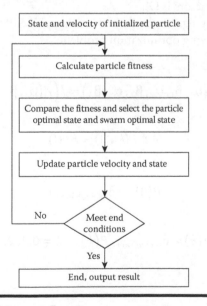

Figure 9.2   Algorithm process flow 2.

Step 3: Compare the fitness of every particle and the optimal position PB_i that every particle globally passes by. If the former is better than the latter, then select the former as the currently optimal position.

Step 4: Compare the fitness of every particle and the optimal position GB that the swarm passes by. If the former is better than the latter, then select the former as the currently optimal position.

Step 5: Update the particle speed and position according to Formula 9.1.

Step 6: If the end conditions are met, output the solution; otherwise, return to Step 2.

9.3 Demonstration Phase

In the demonstration phase, suppose the service life, repair time for repairable maintenance, and logistic delay time follow the probability distributions Weibull(α_1, β_1), Weibull(α_2, β_2), and Weibull(α_3, β_3), respectively, and the truncated points are n_1, n_2, and n_3. Given the inherent availability index A_{i0} and cost constraint C_1, the constraint $(\lambda, \mu, \rho) \in \Theta_1$ on system failure rate, repair rate, and logistic delay rate follows Weibull distribution. Suppose Θ_1 is affected by the demonstration factors (α_1, β_1, α_2, β_2, α_3, β_3). In variable design, α_1 and β_1 reflect the design requirements of the equipment itself, α_2 and β_2 reflect the design requirements on the equipment maintenance subsystem, and α_3 and β_3 reflect the design requirements on the equipment support subsystem. Assume the constraint set of these demonstration factors is Ω_1'.

Then, the system design in the demonstration phase (9.8 through 9.10) can be described by the following optimization problem:

$$\underset{(\alpha_1,\beta_1,\alpha_2,\beta_2,\alpha_3,\beta_3)\in\Omega_1}{\text{OPT}} I\big(\alpha_1,\beta_1,\alpha_2,\beta_2,\alpha_3,\beta_3\big) = I\big(A\big(\alpha_1,\beta_1,\alpha_2,\beta_2,\alpha_3,\beta_3\big)\big) \quad (9.2)$$

$$s.t.\ P(k+1) = BP(k) \quad (9.3)$$

$$P(0) = \big(\delta_{1,n_1+n_2+n_3+2}\big)^T \quad (9.4)$$

$$A(k) = \delta_{n_1+1,n_2+n_3+2}P(k), \quad k = 0, 1, 2, \dots \quad (9.5)$$

where

$$\Omega_1 = \big\{(\alpha_1,\beta_1,\alpha_2,\beta_2,\alpha_3,\beta_3)\in\Omega_1'\big|A_s(\alpha_1,\beta_1,\alpha_2,\beta_2,\alpha_3,\beta_3)\geq A_{i0}\big\}\cap\big\{x\big|C_1(x)\leq C_1\big\}$$

$$
B = \begin{bmatrix}
0 & 0 & \cdots & 0 & 0 & \mu(0) & \mu(1) & \cdots \\
1-\lambda(0) & 0 & \cdots & 0 & 0 & 0 & 0 & \cdots \\
0 & 1-\lambda(1) & \cdots & 0 & 0 & 0 & 0 & \cdots \\
\cdots & \cdots & \cdots & \cdots & \cdots & \cdots & \cdots & \cdots \\
0 & 0 & \cdots & 1-\lambda(n-1_1) & 0 & 0 & 0 & \cdots \\
0 & 0 & \cdots & 0 & 0 & 0 & 0 & \cdots \\
0 & 0 & \cdots & 0 & 0 & 1-\mu(0) & 0 & \cdots \\
0 & 0 & \cdots & 0 & 0 & 0 & 1-\mu(1) & \cdots \\
\cdots & \cdots & \cdots & \cdots & \cdots & \cdots & \cdots & \cdots \\
0 & 0 & \cdots & 0 & 0 & 0 & 0 & \cdots \\
\lambda(0) & \lambda(1) & \cdots & \lambda(n_{-11}) & 1 & 0 & 0 & \cdots \\
0 & 0 & \cdots & 0 & 0 & 0 & 0 & \cdots \\
0 & 0 & \cdots & 0 & 0 & 0 & 0 & \cdots \\
\cdots & \cdots & \cdots & \cdots & \cdots & \cdots & \cdots & \cdots \\
0 & 0 & \cdots & 0 & 0 & 0 & 0 & \cdots
\end{bmatrix}
$$

$$
\begin{bmatrix}
\mu(n_2-1) & 1 & 0 & 0 & \cdots & 0 & 0 \\
0 & 0 & 0 & 0 & \cdots & 0 & 0 \\
0 & 0 & 0 & 0 & \cdots & 0 & 0 \\
\cdots & \cdots & \cdots & \cdots & \cdots & \cdots & \cdots \\
0 & 0 & 0 & 0 & \cdots & 0 & 0 \\
0 & 0 & \rho(0) & \rho(1) & \cdots & \rho(n_3-1) & 1 \\
0 & 0 & 0 & 0 & \cdots & 0 & 0 \\
0 & 0 & 0 & 0 & \cdots & 0 & 0 \\
\cdots & \cdots & \cdots & \cdots & \cdots & \cdots & \cdots \\
1-\mu(n_2-1) & 0 & 0 & 0 & \cdots & 0 & 0 \\
0 & 0 & 0 & 0 & \cdots & 0 & 0 \\
0 & 0 & 1-\rho(0) & 0 & \cdots & 0 & 0 \\
0 & 0 & 0 & 1-\rho(1) & \cdots & 0 & 0 \\
\cdots & \cdots & \cdots & \cdots & \cdots & \cdots & \cdots \\
0 & 0 & 0 & 0 & \cdots & 1-\rho(n_3-1) & 0
\end{bmatrix}
$$

$$
A_s = \frac{1}{D}\left(1 + \sum_{i=2}^{n_1+1}\left\{\prod_{j=0}^{i-2}[1-\lambda(j)]\right\}\right)
$$

$$D = 3 + \sum_{i=2}^{n_1+1}\left\{\prod_{j=0}^{i-2}\left[1-\lambda(j)\right]\right\} + \sum_{i=2}^{n_2+1}\left\{\prod_{j=0}^{i-2}\left[1-\mu(j)\right]\right\} + \sum_{i=2}^{n_3+1}\left\{\prod_{j=0}^{i-2}\left[1-\rho(j)\right]\right\}$$

$$\lambda(k) = 1 - \alpha_1^{(k+1)^{\beta_1}-k^{\beta_1}}, \quad k = 0, 1, 2, \ldots n_1 - 1, \quad \lambda(n_1) = 1$$

$$\mu(k) = 1 - \alpha_2^{(k+1)^{\beta_2}-k^{\beta_2}}, \quad k = 0, 1, 2, \ldots n_2 - 1, \quad \mu(n_2) = 1$$

$$\rho(k) = 1 - \alpha_3^{(k+1)^{\beta_3}-k^{\beta_3}}, \quad k = 0, 1, 2, \ldots n_3 - 1, \quad \rho(n_3) = 1.$$

9.3.1 Design of System Minimum Availability Amplitude

Generally, any new equipment systems want minimized fluctuation in instantaneous availability in order to maintain the steady-state availability, even though the availability amplitude decreases. Therefore, the minimum availability amplitude design can be described by the following optimization model:

$$\min_{(\alpha_1, \beta_1, \alpha_2, \beta_2, \alpha_3, \beta_3) \in \Omega_1} M(\alpha_1, \beta_1, \alpha_2, \beta_2, \alpha_3, \beta_3) = A - \min_{k}\{A(k)\} \qquad (9.6)$$

s.t. Formulas 9.3 through 9.5 are satisfied

9.3.2 Optimal System Matching Design

The physical meaning of system matching means that the maintenance subsystem and support subsystem can coordinate with the equipment itself to always maintain system instantaneous availability above a certain level. The system adaptation time to some extent reflects the convergence time from instantaneous availability to steady-state availability. The matched system is the system with an adaptation time at 0. Its optimal matching design, is to minimize its adaptation with the properly selected design variables; this is called the minimized adaptation time design of the system.

This can be described as a constrained six-variable optimization problem.

$$\min_{(\alpha_1, \beta_1, \alpha_2, \beta_2, \alpha_3, \beta_3) \in \Omega_1} T(\alpha_1, \beta_1, \alpha_2, \beta_2, \alpha_3, \beta_3) = T\left(A(\alpha_1, \beta_1, \alpha_2, \beta_2, \alpha_3, \beta_3)\right) \qquad (9.7)$$

s.t. Formulas 9.3 through 9.5 are satisfied

All that is needed is to adjust Formula 9.6 accordingly based upon the research on other fluctuation parameters.

9.3.3 Simulation Experiment

Following is the sample analysis on models (9.6 and 9.7) via PSO with real-value encoding. The experimental environment included a PC computer, Celeron(R) 1.80 GHz CPU; 0.99 GB RAM; Windows XP; simulation software MATLAB® 7.3; particle swarm size 40, maximum evolution generation 100, acceleration constants $c_1 = c_2 = 1.4962$ [104], and inertia weight $\omega = 0.7298$. The environment for simulation experiments that follow are all identical to this example.

Example 1

Suppose

$$\Omega_1 = \{(\alpha_1, \beta_1, \alpha_2, \beta_2, \alpha_3, \beta_3) \mid \alpha_{11} \le \alpha_1 \le \alpha_{12}, \beta_{11} \le \beta_1 \le \beta_{12}, \alpha_{21} \le \alpha_2 \le \alpha_{22},$$
$$\beta_{21} \le \beta_2 \le \beta_{22}, \alpha_{31} \le \alpha_3 \le \alpha_{32}, \beta_{31} \le \beta_3 \le \beta_{32}\}.$$

The constraints are set as follows:

$\alpha_{11} = 0.95, \quad \alpha_{12} = 0.99998, \quad \beta_{11} = 0.95, \quad \beta_{12} = 4$

$\alpha_{21} = 0.9, \quad \alpha_{22} = 0.9998, \quad \beta_{21} = 0.95, \quad \beta_{22} = 3$

$\alpha_{31} = 0.9, \quad \alpha_{32} = 0.9995, \quad \beta_{31} = 0.95, \quad \beta_{32} = 3;$

the truncated points are $n_1 = n_2 = n_3 = 200$, and the research time interval for the system is $N = 600$. In order to describe the fluctuation characteristics of system instantaneous availability in the case of a fixed steady-state index for system instantaneous availability–system steady-state availability, the constraint for this steady-state availability example is $A = 0.75$. To simplify—none of the examples in this chapter will consider the cost and other expenses, without loss of generality. The cost can be a constraint added to Model 9.6 and Model 9.7.

For the design of system minimum availability amplitude (9.6), the particle fitness can be defined as

$$Fitness = \begin{cases} 1 + (A - 0.75)^2 & |A - 0.75| > 0.001 \\ M & |A - 0.75| \le 0.001. \end{cases}$$

With the PSO algorithm mentioned in 9.2.1, the optimal approximate solutions can be obtained for problem (9.6):

$$\alpha_1^* = 0.97404600744796, \quad \beta_1^* = 1.18410316863168$$

$$\alpha_2^* = 0.93134004794793, \quad \beta_2^* = 2.47185280112372$$

$$\alpha_3^* = 0.96432511721495, \quad \beta_3^* = 2.49531088692593.$$

The optimal approximate is value is

$$M^* = 8.850344845878766\text{e-}007.$$

The corresponding system instantaneous availability is shown in Figure 9.3.

To clarify, the optimal approximate design result is compared with another feasible design result. Suppose the design variables for system 1 are optimal approximate solutions $\alpha_1^*, \beta_1^*, \alpha_2^*, \beta_2^*, \alpha_3^*,$ and β_3^*; and the design variables for system 2 in feasible design are

$$\alpha_1 = 0.99995000000000, \quad \beta_1 = 3.16899999999999$$
$$\alpha_2 = 0.93134004794793, \quad \beta_2 = 2.47185280112372$$
$$\alpha_3 = 0.96432511721495, \quad \beta_3 = 2.49531088692593.$$

The comparison of instantaneous availability for these two systems is shown in Figure 9.4.

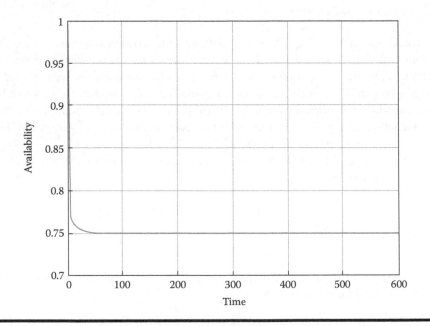

Figure 9.3 System instantaneous availability curve corresponding to optimal approximate solution of minimum availability amplitude.

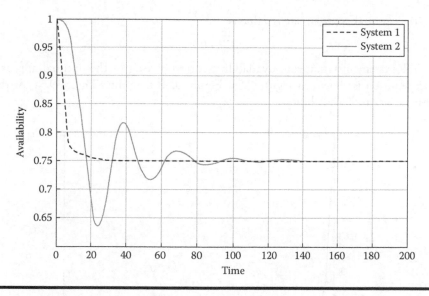

Figure 9.4 Comparison of instantaneous availability for two systems.

As seen from Figure 9.4, the instantaneous availability of system 1 decreases monotonically and tends toward a steady state; whereas the instantaneous availability of system 2 fluctuates heavily at an early stage, with the availability amplitude exceeding 0.2. It then takes about 140 units of time to stabilize at the steady state. In this example, both systems meet the design requirements and have the same steady-state availability. But considering the characteristics of instantaneous availability fluctuation, system 1 is obviously better than system 2. The design parameter for system 1 is obtained by PSO, which makes the amplitude of system instantaneous availability almost 0. This means a design set has been found within the constrained range to meet the system matching design.

For the optimal system matching design (9.7), the availability standardized level is set at $\varepsilon_0 = 0.005$; then the particle fitness can be defined as

$$Fitness = \begin{cases} N + (A - 0.75)^2 & |A - 0.75| > 0.001 \\ T & |A - 0.75| \le 0.001. \end{cases}$$

We use PSO in 6.2.1 to find the optimal approximate solutions as follows:

$$\alpha_1^* = 0.99388755485497, \quad \beta_1^* = 1.40877172125705$$

$$\alpha_2^* = 0.90489072744946, \quad \beta_2^* = 1.21816646480170$$

$$\alpha_3^* = 0.98348893754253, \quad \beta_3^* = 2.61451071435671.$$

The optimal approximate is value is

$$T^* = 0.$$

The system instantaneous availability corresponding to the optimal approximate solution is shown in Figure 9.5a. Figure 9.5b magnifies Figure 9.5a for the great variation of availability.

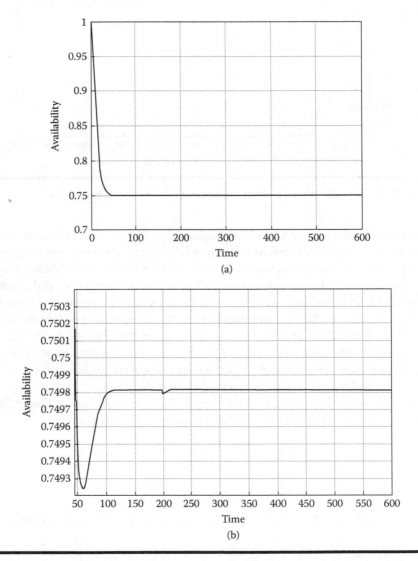

(a)

(b)

Figure 9.5   (a) System instantaneous availability curve corresponding to optimal approximate solution with minimum adaptation time. (b) System instantaneous availability curve corresponding to optimal approximate solution with minimum adaptation time.

The system adaptation time corresponding to the optimal approximate solution is 0, which means the system is matched. As seen from Figures 9.4 and 9.5, the system instantaneous availability almost monotonically decreases and tends to stabilize at the steady state. But it fluctuates at [50,100], and also fluctuates slightly at 200th unit time. The fluctuation at [50,100] refers to the maintenance subsystem of the equipment caused by the coordination between the support subsystem and the equipment itself. The fluctuation at 200th unit time is caused by the truncated point under truncated and discrete Weibull distribution also at 200th unit time. But these fluctuations are completely below the availability standardized level; the system is still matched.

Through the comparison between the design results of minimum availability amplitude and minimum adaptation time, there is a big difference in their design variables, but they all make the system matched, which means the system matching design is not unique. It can vary with the specific conditions in the actual engineering practices. In this example, the design of the minimum availability amplitude and minimum adaptation time makes the system matched, which is related to the constraints. In real equipment design, sometimes the constraints are extremely strict so that any feasible design cannot make the system matched. In these cases, it is more meaningful to solve the minimum adaptation time of the system, which will occur in the examples that follow.

9.4 Development Phase

Suppose the system service life and repair time follow the probability distributions Weibull(α_1, β_1) and Weibull(α_2, β_2), respectively, and the truncated points are n_1 and n_2 given the system mean time between failures $MTBF_0$, mean time to repair $MTTR_0$, and cost constraint C_2. Constraint $(\lambda, \mu) \in \Theta_2$ on system failure rate and repair rate follows Weibull distribution, and suppose Θ_2 is affected by the demonstration factors $(\alpha_1, \beta_1, \alpha_2, \beta_2)$. Assume the constraint set of these demonstration factors is Ω_2'. The system design in the development phase can be described as the following optimization problem:

$$\underset{(\alpha_1, \beta_1, \alpha_2, \beta_2) \in \Omega_2}{\text{OPT}} \quad I(\alpha_1, \beta_1, \alpha_2, \beta_2) = I\big(A(\alpha_1, \beta_1, \alpha_2, \beta_2)\big) \tag{9.8}$$

$$s.t. \ P(k+1) = BP(k) \tag{9.9}$$

$$P(0) = \left(\delta_{1,n_1+n_2+1}\right)^T \tag{9.10}$$

$$A(k) = \delta_{n_1+1,n_2+1} P(k), \quad k = 0, 1, 2, \dots \tag{9.11}$$

where

$$\Omega_2 = \left\{ (\alpha_1, \beta_1, \alpha_2, \beta_2) \in \Omega_1' \middle| MTBF\left(\lambda(\alpha_1, \beta_1)\right) \geq MTBF_0, \right.$$
$$\left. MTTR\left(\mu(\alpha_2, \beta_2)\right) \leq MTTR_0 \right\} \cap \left\{ x \middle| C_2(x) \leq C_2 \right\}$$

$$B = \begin{bmatrix}
0 & 0 & \cdots & 0 & 0 & \mu(0) & \mu(1) & \cdots & \mu(n_2-1) & 1 \\
1-\lambda(0) & 0 & \cdots & 0 & 0 & 0 & 0 & \cdots & 0 & 0 \\
0 & 1-\lambda(1) & \cdots & 0 & 0 & 0 & 0 & \cdots & 0 & 0 \\
\cdots & \cdots & \cdots & \cdots & \cdots & \cdots & \cdots & \cdots & \cdots & \cdots \\
0 & 0 & \cdots & 1-\lambda(n_1-1) & 0 & 0 & 0 & \cdots & 0 & 0 \\
\lambda(0) & \lambda(1) & \cdots & \lambda(n_1-1) & 1 & 0 & 0 & \cdots & 0 & 0 \\
0 & 0 & \cdots & 0 & 0 & 1-\mu(0) & 0 & \cdots & 0 & 0 \\
0 & 0 & \cdots & 0 & 0 & 0 & 1-\mu(1) & \cdots & 0 & 0 \\
\cdots & \cdots & \cdots & \cdots & \cdots & \cdots & \cdots & \cdots & \cdots & \cdots \\
0 & 0 & \cdots & 0 & 0 & 0 & 0 & \cdots & 1-\mu(n_2-1) & 0
\end{bmatrix}$$

$$A_s = \frac{1}{D}\left(1 + \sum_{i=2}^{n_1+1}\left\{ \prod_{j=0}^{i-2}\left[1-\lambda(j)\right] \right\} \right)$$

$$D = 2 + \sum_{i=2}^{n_1+1}\left\{ \prod_{j=0}^{i-2}\left[1-\lambda(j)\right] \right\} + \sum_{i=2}^{n_2+1}\left\{ \prod_{j=0}^{i-2}\left[1-\mu(j)\right] \right\}$$

$$\lambda(k) = 1 - \alpha_1^{(k+1)^{\beta_1} - k^{\beta_1}}, \quad k = 0, 1, 2, \ldots, n_1 - 1, \quad \lambda(n_1) = 1$$

$$\mu(k) = 1 - \alpha_2^{(k+1)^{\beta_2} - k^{\beta_2}}, \quad k = 0, 1, 2, \ldots, n_2 - 1, \quad \mu(n_2) = 1.$$

9.4.1 Optimal System Matching Design

This can be described as a constrained four-variable optimization problem, such as 9.2.3:

$$\underset{(\alpha_1, \beta_1, \alpha_2, \beta_2) \in \Omega_2}{\text{OPT}} \quad T(\alpha_1, \beta_1, \alpha_2, \beta_2) = T(A(\alpha_1, \beta_1, \alpha_2, \beta_2)) \qquad (9.12)$$

s.t. Formulas 9.9 through 9.11 are satisfied

9.4.2 Simulation Experiment

Example 2

Suppose

$$\Omega_1 = \{(\alpha_1, \beta_1, \alpha_2, \beta_2) \mid \alpha_{11} \le \alpha_1 \le \alpha_{12}, \beta_{11} \le \beta_1 \le \beta_{12}, \alpha_{21} \le \alpha_2 \le \alpha_{22}, \beta_{21} \le \beta_2 \le \beta_{22}\}.$$

The constraints are set as follows:

$$\alpha_{11} = 0.8, \quad \alpha_{12} = 0.9999, \quad \beta_{11} = 0.7, \quad \beta_{12} = 4$$

$$\alpha_{21} = 0.995, \quad \alpha_{22} = 0.9998, \quad \beta_{21} = 3, \quad \beta_{22} = 4.$$

The truncated points are $n_1 = n_2 = 200$, and the research time interval for the system is $N = 600$. In order to describe the fluctuation characteristics of system instantaneous availability in the case of fixed steady-state indexes for system instantaneous availability $MTBF$ and $MTTR$, the constraints are set as $MTBF = 9.43$ and $MTTR = 5.72$.

For the design of system minimum availability amplitude (9.12), the particle fitness can be defined as

$$Fitness = \begin{cases} N + (MTBF - 9.43)^2 + (MTTR - 5.72)^2 & |MTBF - 9.43| + |MTTR - 5.72| > 0.01 \\ T & |MTBF - 9.43| + |MTTR - 5.72| \le 0.01. \end{cases}$$

With the PSO algorithm mentioned in 9.2.1, the optimal approximate solution can be obtained for problem (9.12):

$$\alpha_1^* = 0.87148012093951, \quad \beta_1^* = 0.92200000000011$$

$$\alpha_2^* = 0.99833135796469, \quad \beta_2^* = 3.64299999999982.$$

The optimal approximate is value is

$$T^* = 9.$$

The corresponding system instantaneous availability is shown in Figure 9.6.

To be clearer, the optimal approximate design result is compared with another feasible design result. The variables in the feasible design are

$$\alpha_1 = 0.83998000000000, \quad \beta_1 = 0.83500000000011$$

$$\alpha_2 = 0.99740000000000, \quad \beta_2 = 3.38199999999985.$$

The system adaptation time corresponding to this feasible design is

$$T = 18.$$

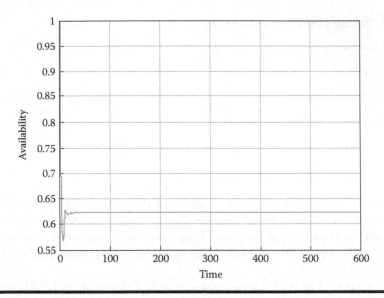

Figure 9.6 **System instantaneous availability curve 1 corresponding to optimal approximate solution for optimal matching.**

The comparison between the instantaneous availability from the feasible design and the instantaneous availability from the optimal approximate design is shown in Figure 9.7. From the adaptation time view, the optimal design is obviously better than this feasible design.

As seen from Figure 9.7, with the parameters α_1, β_1, and *MTTR* fixed under the service life distribution, as the parameter β_2 under repair time distribution increases, the corresponding system adaptation time also increases; therefore, the optimal design value β_1 should be the minimum value. It can be also noted that if β_2 is adjusted above a certain critical value, then the system adaptation time almost remains unchanged for this adjustment. Comparing optimization result and rules obtained from Chapter 8 with this example, it can be considered that the critical value for adjusting β_2 according to the system adaptation time should be lower than the infimum 3 for β_2 design range. As seen in Figure 8.3, with the parameters α_2, β_2, and *MTBF* fixed under system repair time distribution, the system adaptation time has the minimum value near $\beta_1 = 1$, as reflected in the optimization result.

Actually the rules analyzed from Chapter 8 can further simplify this optimization problem. In view of this, the optimal values of two parameters in the design variables should be

$$\beta_2^* = \min\left\{\beta_2 \middle| MTTR(\alpha_2, \beta_2) = 5.72, 0.995 \leq \alpha_2 \leq 0.9998, 3 \leq \beta_2 \leq 4\right\}$$

$$\alpha_2^* \in \left\{\alpha_2 \middle| MTTR\left(\alpha_2, \beta_2^*\right) = 5.72\right\}.$$

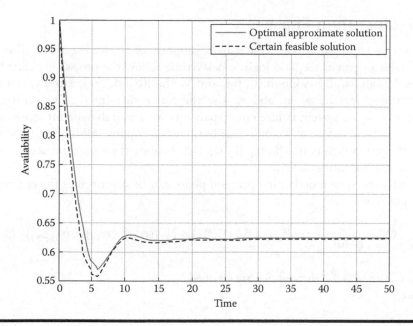

Figure 9.7 Comparison of system instantaneous availability corresponding to two different designs.

Suppose

$$\alpha_1^* = 0.99500576000000$$

$$\beta_1^* = 3.00099999999989.$$

Now the four-parameter optimization is degraded to a two-parameter optimization in this example. Optimize the left two parameters and obtain the approximate solutions as follows:

$$\alpha_1^* = 0.88384459984087$$

$$\beta_1^* = 0.96300000000011.$$

The corresponding system adaptation time is

$$T = 10.$$

The solutions from these two methods are close to each other. Therefore, both methods can be used to solve the models, but the conclusions from Chapter 8 can simplify the solution process.

9.5 Operation Phase

In the case of no preventive maintenance, suppose the service life, repair time for repairable maintenance, and logistic delay time follow the probability distributions Weibull(α_1, β_1), Weibull(α_2, β_2), and Weibull(α_3, β_3), respectively, and the truncated points are n_1, n_2, and n_3. Use the cost constraint C_3. The constraint $(\lambda, \mu, \rho) \in \Theta_3$ on system failure rate, repair rate for repairable maintenance, and logistic delay rate follow Weibull distribution, and suppose Θ_3 is affected by the demonstration factors $(\alpha_1, \beta_1, \alpha_2, \beta_2, \alpha_3, \beta_3)$. Assume the constraint set of these demonstration factors is Ω'_3.

Then, the system design in operation phase can be described as the optimization problem:

$$\underset{(\alpha_1, \beta_1, \alpha_2, \beta_2, \alpha_3, \beta_3) \in \Omega_3}{\text{OPT}} I(\alpha_1, \beta_1, \alpha_2, \beta_2, \alpha_3, \beta_3) = I\big(A(\alpha_1, \beta_1, \alpha_2, \beta_2, \alpha_3, \beta_3)\big) \quad (9.13)$$

s.t. Formulas 9.2 through 9.4 are satisfied
where

$$\Omega_3 = \Omega'_3 \cap \{x | C_3(x) \leq C_3\}.$$

The design in operation phase (9.13) is similar to that in demonstration phase (9.2). The difference lies in the constraints. But for analyzing the optimization algorithm, no difference is considered between them, so this is no longer repeated.

Considering the preventive maintenance, suppose the system service life, repair rate for repairable maintenance, and repair time for preventive maintenance follow the probability distributions Weibull(α_1, β_1), Weibull(α_2, β_2), and Weibull(α_3, β_3), respectively, and set $n_0(<n_1)$ as the preventive maintenance cycle. Then the system design in operation phase (9.3 through 9.6) can be described as following the optimization problem:

$$\underset{(\alpha_1, \beta_1, \alpha_2, \beta_2, \alpha_3, \beta_3) \in \Omega, n_0 \in Z^+}{\text{OPT}} I(\alpha_1, \beta_1, \alpha_2, \beta_2, \alpha_3, \beta_3, n_0) = I(A(\alpha_1, \beta_1, \alpha_2, \beta_2, \alpha_3, \beta_3, n_0))$$

$$(9.14)$$

$$\textit{s.t. } P(k + 1) = BP(k) \quad (9.15)$$

$$P(0) = (\delta_{1, n_1 + n_2 + n_3 + 2})^T \quad (9.16)$$

$$A(k) = \delta_{m+1, n_2 + n_3 + 2} p(k), \quad k = 0, 1, 2, \dots \quad (9.17)$$

where

$$
B = \begin{bmatrix}
0 & 0 & \cdots & 0 & 0 & \mu_1(0) & \mu_1(0) & \cdots \\
1-\lambda(1) & 0 & \cdots & 0 & 0 & 0 & 0 & \cdots \\
0 & 1-\lambda(1) & \cdots & 0 & 0 & 0 & 0 & \cdots \\
\cdots & \cdots & \cdots & \cdots & \cdots & \cdots & \cdots & \cdots \\
0 & 0 & \cdots & 1-\lambda(n_0-1) & 0 & 0 & 0 & \cdots \\
\lambda(0) & \lambda(1) & \cdots & \lambda(n_0-1) & \lambda(n_0) & 0 & 0 & \cdots \\
0 & 0 & \cdots & 0 & 0 & 1-\mu_1(0) & 0 & \cdots \\
0 & 0 & \cdots & 0 & 0 & 0 & 1-\mu_1(1) & \cdots \\
\cdots & \cdots & \cdots & \cdots & \cdots & \cdots & \cdots & \cdots \\
0 & 0 & \cdots & 0 & 0 & 0 & 0 & \cdots \\
0 & 0 & \cdots & 0 & 1-\lambda(n_0) & 0 & 0 & \cdots \\
0 & 0 & \cdots & 0 & 0 & 0 & 0 & \cdots \\
0 & 0 & \cdots & 0 & 0 & 0 & 0 & \cdots \\
\cdots & \cdots & \cdots & \cdots & \cdots & \cdots & \cdots & \cdots \\
0 & 0 & \cdots & 0 & 0 & 0 & 0 & \cdots
\end{bmatrix}
$$

$$
\begin{bmatrix}
\mu_1(n_2-1) & 1 & \mu_2(0) & \mu_2(1) & \cdots & \mu_2(n_3-1) & 1 \\
0 & 0 & 0 & 0 & \cdots & 0 & 0 \\
0 & 0 & 0 & 0 & \cdots & 0 & 0 \\
\cdots & \cdots & \cdots & \cdots & \cdots & \cdots & \cdots \\
0 & 0 & 0 & 0 & \cdots & 0 & 0 \\
0 & 0 & 0 & 0 & \cdots & 0 & 0 \\
0 & 0 & 0 & 0 & \cdots & 0 & 0 \\
0 & 0 & 0 & 0 & \cdots & 0 & 0 \\
\cdots & \cdots & \cdots & \cdots & \cdots & \cdots & \cdots \\
1-\mu_1(n_2-1) & 0 & 0 & 0 & \cdots & 0 & 0 \\
0 & 0 & 0 & 0 & \cdots & 0 & 0 \\
0 & 0 & 1-\mu_2(0) & 0 & \cdots & 0 & 0 \\
0 & 0 & 0 & 1-\mu_2(1) & \cdots & 0 & 0 \\
\cdots & \cdots & \cdots & \cdots & \cdots & \cdots & \cdots \\
0 & 0 & 0 & 0 & \cdots & 1-\mu_2(n_3-1) & 0
\end{bmatrix}.
$$

Here the mean availability or the interval availability within the equipment service period is selected as the index for research. As seen from Figure 8.16 in Chapter 8, with α_1, β_1, α_2, β_2 and *MPMT* fixed, as the parameter under preventive maintenance time distribution decreases, the system interval availability decreases, directly selecting the infimum of β_3 in the design. The optimal design of α_1, β_1, α_2, β_2 can be solved similarly by PSO mentioned earlier. This is mainly to study the commonly used optimal preventive maintenance cycle model.

9.5.1 Optimal Preventive Maintenance Cycle

When the system service life, repairable maintenance, and the repair time distribution for repairable maintenance are known, namely $(\alpha_1, \beta_1, \alpha_2, \beta_2, \alpha_3, \beta_3)$ are given, the optimal preventive maintenance cycle model can be mathematically described by the following optimization problem:

$$\underset{n_0}{\text{OPT}} \ \bar{A}(n_0) = \frac{1}{N+1} \sum_{i=0}^{N} A(i) \tag{9.18}$$

s.t. Formulas 9.15 through 9.17 are satisfied
where N is the equipment service period.

9.5.2 Simulation Experiment

Example 3

Suppose $\alpha_1 = 0.9998$, $\beta_1 = 2$, $\alpha_2 = 0.999$, $\alpha_3 = 2$, and $\beta_3 = 3$, truncated points are $n_1 = n_2 = n_3 = 200$, and the research time interval for the system $N = 600$. Because Model 6.14 is a one-dimensional integer optimization model, the optimal solution can be obtained by direct point-by-point search:

$$n_0^* = 37.$$

The system interval availability corresponding to different preventive maintenance cycles is shown in Figure 9.8.

As seen from Figure 9.8, when the system has frequent preventive maintenance, it brings a lot of nonfailure down-time, which greatly reduces the system interval availability. However, without the preventive maintenance, the system cannot be well protected from the failures, developing more serious potential hazards. Hence, proper preventive maintenance can greatly improve the system interval availability.

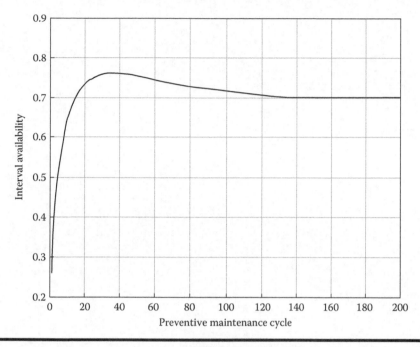

Figure 9.8 System interval availability corresponding to different preventive maintenance cycles.

9.6 Summary

Using truncated Weibull distribution conditions, the research in this chapter combines the rules obtained from the simulation analysis in Chapter 8 and studies the constrained multivariable optimization models describing the fluctuation characteristic parameters of system instantaneous availability from the demonstration, development, and operation phases within the life cycle. Then, PSO is used as the optimization tool to simulate and illustrate the model's effectiveness.

Chapter 10

Conclusions and Prospects

This book mainly studies the matching transition of new equipment systems as a result of interactions among subsystems, which include reliability engineering, maintainability engineering, supportability engineering, and engineering cybernetics. The main idea of this book is to further extend the research on steady-state availability analysis for optimization and control over instantaneous availability that is more suitable for engineering practices. This extensive research includes

1. Analysis and study of the instantaneous availability models for the existing single-unit repairable systems, repairable systems with repair delay, and repairable systems with preventive maintenance, to set up the instantaneous availability models constrained by limited periods, to prove the existence of system steady-state availability with the theory and methods of matrices, and to obtain the expression of steady-state availability.

2. Because the instantaneous availability of typical repairable systems is in a steady state, the research herein mainly focuses on variations before the instantaneous availability becomes steady, that is, the fluctuation (variation) characteristics of system instantaneous availability within a limited period. Through analysis, a fluctuation parameter system was set up describing the fluctuation characteristics of system instantaneous availability within a limited time, which well reflects the fluctuation level of system instantaneous availability to some extent. And we also set up the optimal control models for these parameters.

3. Conducting the simulation analysis on the instantaneous availability models for the repairable systems under truncated Weibull distribution; in cases

271

where the steady-state indexes of systems remain unchanged—such as the mean time to repair, mean time between failures, or mean logistic delay time—we studied the variation of parameters describing the fluctuation of system instantaneous availability with respect to the characteristic parameters under related time distribution, and then we summarized some rules.

4. Under truncated Weibull distribution, the optimal control model based on system matching was degraded to the constrained multi-variable optimal model. From the demonstration, development, and operation phase within the equipment whole service life cycle, this book studies the optimization models for minimum availability, optimal system matching, and optimal preventive maintenance cycle. Finally, the Particle Swarm Optimization (PSO) is used as the optimization tool to simulate and illustrate the model effectiveness.

The research herein shows that, no matter what the theoretical study and engineering applications, instantaneous availability is a very difficult problem. This book only shows the tip of the iceberg. There is still a lot to be solved in the future, specifically including

1. To further supplement and complete the fluctuation characteristic parameters of instantaneous availability within limited periods, to further improve the research on system optimization of instantaneous availability, and to try to find out the basic principles in the coordination among the failure time variation rule, repair time variation rule, and logistic delay time rule, to create conditions for instructing the engineering practices.

2. The fluctuation of equipment availability is affected by its own RMS characteristics of which the influencing factors vary with the equipment types in different phases within the equipment's whole service life cycle. It is very important to extract and study the specific factors in each phase; this will be proposed in future research for different equipment types and different phases.

3. The equipment subsystem mainly consists of two elements, showing a structural relationship among equipment units. This book mainly studies the inter-matching relations among maintenance subsystems, support subsystems, and the equipment itself. But the qualitative and quantitative relations between the internal structure of equipment subsystems and the instantaneous availability fluctuation of the equipment still need to be further studied.

Addressing these problems is critical for solving the problems outlined in this book. However, we have provided only general mathematical models described in abstract concepts, which provide an overall theoretical framework for the next intensive research.

References

1. Macheret Y, Koehn P and Sparrow D. Improving reliability and operational availability of military systems. *2005 IEEE Aerospace Conference.* Big sky, MT: IEEE, 2005, 3489–3957.
2. Ma Shaomin. *Comprehensive protection engineering.* Beijing: Defense Industry Press, 1995.
3. Kang Rui and Yu Yong-Li. Theory and practice of equipment reliability, maintainability and supportability engineering. *China Mechanical Engineering,* 1998, 9(12): 3–6.
4. Yang Weimin, Yuan Lian and Tu Qingci. Reliability systems engineering–theory and practice. *Acta Aeronautica Et Astronautica Sinica,* 1995, 16(Suppl 1): 1–8.
5. Yang Weimin. *Overview on reliability, maintainability and supportability.* Beijing: National Defense Industry Press, 1995.
6. Kang Rui and Wang Zili. Theory and technical framework of reliability system engineering. *Acta Aeronautica Et Astronautica Sinica,* 2005, 26(5): 633–636.
7. Yang Weimin and Tu Qingci. 21st century equipment reliability, maintainability and supportability engineering development framework study. *Chinese Mechanical Engineering,* 1998, 9(12): 45–48.
8. Cao Jinhua and Cheng Kan. *Reliability mathematics introduction.* Beijing: Beijing Science Press, 1986.
9. Zhao Yuejin and Dongwei Yan. A new algorithm and discovery for comprehensive system inherent availability. *Journal of Beijing Institute of Technology,* 2005, 25(5): 54–57.
10. Kong Deliang and Wang Shaoping. Availability analysis methods to repairable system. *Journal of Beijing University of Aeronautics and Astronautics,* 2002, 28(2): 129–132.
11. Zeng Shengkui, Zhao Tingdi and Zhang Jianguo, et al. *System reliability design and analysis tutorial.* Beijing: Beijing University of Aeronautics and Astronautics Press, 2004.
12. Xiao Gang. A Monte Carlo method for obtaining reliability and availability confidence limits of complex maintenance system. *Acta Armamentarii,* 2002, 23(2): 215–218.
13. Zheng Z H, Cui L R and Hawkes A G. A study on a single-unit Markov repairable system with repair time omission. *IEEE Transactions on Reliability,* 2006, 55(5): 182–188.
14. Hassett T F, Dietrich D L and Szidarovszky F. Time-varying failure rates in the availability & reliability analysis of repairable systems. *IEEE Transactions on Reliability,* 1995, 44(1): 155–160.

15. Sun H R and Han J J. Instantaneous availability and interval availability for systems with time-varying failure rate: stair-step approximation. *Proceedings of 2001 Pacific Rim International Symposium on Dependable Computing*, Seoul: IEEE, 2001, 371–374.

16. Zhang T and Horigome M. Availability and reliability of system with dependent components and time-varying failure and repair rates. *IEEE Transactions on Reliability*, 2001, 50(2): 151–158.

17. Sarkar J and Chandhuri G. Availability of a system with gamma life and exponential repair time under a perfect repair policy. *Statistics & Probability Letters*, 1999, 43: 189–196.

18. Cassady C R, Lyoob I M and Schneider K, et al. A generic model of equipment availability under imperfect maintenance. *IEEE Transactions on Reliability*, 2005, 54(4): 564–571.

19. Mi J. Limiting availability of system with non-identical lifetime distributions and non-identical repair time distributions. *Statistics & Probability Letters*, 2006, 76(7): 729–736.

20. Biswas A and Sarkar J. Availability of a system maintained through several imperfect repairs before a replacement or a perfect repair. *Statistics & Probability Letters*, 2000, 50(2): 105–114.

21. Zhang Yingfeng and Jiang Tao. Analysis to repairable system steady state availability with Markov process. *Journal of Xuzhou Normal University*, 2000, 18(2): 27–28.

22. Bris R, Chatelet E and Yalaoui F. New method to minimize the preventive maintenance cost of series-parallel systems. *Reliability Engineering & System Safety*, 2003, 82(3): 247–255.

23. Li Jiangtao, Jiang Liqiang and Huang Lipo. Best equipment preventive maintenance interval study. *Journal of the Academy of Equipment Command & Technology*, 2004, 15(3): 26–29.

24. Zheng Zhihua. Faults impacts on ignored or delayed single components Markov repairable system research. Master's Thesis, Beijing: Beijing Institute of Technology, 2006.

25. Sarkar J and Sarkar S. Availability of a periodically inspected system supported by a spare unit, under perfect repair or perfect upgrade. *Statistics & Probability Letters*, 2001, 53(2): 207–217.

26. Cui L R and Xie M. Availability of a periodically inspected system with random repair or replacement times. *Journal of Statistical Planning and Inference*, 2005, 131(1): 89–100.

27. Kijima M, Morimura H and Suzuki Y. Periodic replacement problem without assuming minimal repair. *European Journal of Operational Research*, 1988, 37(2): 194–203.

28. Kijima M. Some results for repairable system with general repair. *Journal of Applied Probability*, 1989, 26(1): 89–102.

29. Pham H and Wang H. Imperfect maintenance. *European Journal of Operational Research*, 1996, 94(3): 425–438.

30. Zhang Yuanlin and Lin Huineng. Repairable system with Markov dependence and maintenance priority linear adjacent 2/n (F). *Acta Automatica Sinica*, 2000, 26(3): 317–323.

31. Sun P and Meng L. Reliability of a consecutive-k-out-of-n: F system of Markov dependent components. *IEEE Transactions on Reliability*, 1987, 36(1): 76–79.

32. Ber S M. A bivariate Markov process with diffusion and discrete components. *Stochastic Models*, 1994, 10: 271–308.
33. Cheng Kan and Cao Jinhua. General repairable system reliability analysis- Markov update model. *Acta Mathematicae Applicatae Sinica*, 1981, 4: 295–306.
34. Shi Dinghua and He Daojie. Application of vector Markov process approach in repairable systems. *Journal of Shanghai University*, 1996, 2(3): 249–257.
35. Li Cailiang, Pu Bingyuan and Tang Yinghui, et al. Reliability analysis to two different components cold supply system. *Journal of Electronic Science and Technology*, 2003, 32(4): 447–550.
36. Meyn S P and Tweedie R L. *Markov chains and stochastic stability*. London: Springer, 1993.
37. Ohaki S. System reliability analysis by Markov renewal processes. *Journal of the Operational research society of Japan*, 1970, 12: 127–188.
38. Xie Wenxiang, Ru Feng and Xue Junyi. Redundant repairable system availability stochastic petri net modeling and analysis. *Journal of Systems Engineering*, 1998, 13(4): 93–97.
39. Xie Wenxiang and Xue Junyi. Availability calculation based on stochastic Petri nets repairable control system. *Journal of Xi'an Jiaotong University*, 1997, 31(11): 83–87, 116.
40. Ormon Stephen W, Cassady C Richard and Greenwood A G. A simulation based reliability prediction model for conceptual design. *IEEE 2001 Proceedinigs Annual Reliability and Maintainability and Mainatainability Symposium*, Philadelphia, PA: IEEE, 2001, 433–436.
41. Ofelia Gonzalez-Vega, Joseph W Foster III and Hogg G L. A simulation program to model effects of logistics on R&M of complex system. *IEEE Proceedings Annual Reliability and Maintainability Symposium*, Los Angeles, CA: IEEE, 1998, 306–313.
42. Bi Hongkui, Wang Hong and Huang Shujun. Radar system availability computer simulation calculation. *Modern Radar*, 2004, 26(2): 4–5.
43. Gao Wen, Zhu Mingfa and Xu Zhiwei. Cluster system availability Simulation algorithm based on repair time constraints. *Chinese Journal of Computers*, 2001, 24(8): 876–880.
44. Ma Haifeng. Ways to improve the availability of aircraft. *Aircraft Design*, 2001, 4: 37–39.
45. Jiang Chaoyi. Design and implementation of combat unit availability and mission effectiveness simulation evaluation system. Master's Thesis, Ordnance Engineering College, Shijiazhuang, 2004.
46. Upadhya K Sadananda and Srinivasan N K. Availability of weapon systems with logistic delays: a simulation approach. *International Journal of Reliability, Quality and Safety Engineering*, 2003, 10(4): 429–443.
47. Upadhya K Sadananda and Srinivasan N K. Availability of weapon systems with multiple failures and logistic delays. *International Journal of Quality & Reliability Management*, 2003, 20(7): 836–846.
48. Guo Weihua, Xu Genqi and Xu Houbao. Two different components parallel repairable system stability. *Acta Analysis Functionalis Applicata*, 2003, 5(3): 281–288.
49. Guo Weihua, Xu Houbao and Zhu Guangtian. Asymptotic stability of two-part parallel maintenance system solutions. *System Engineering—Theory & Practice*, 2006, 12: 62–68.

50. Xu Houbao, Guo Weihua and Yu Jingyuan. Steady-state solution to a series repairable system. *Acta Mathematicae Applicatae Sinica*, 2006, 29(1): 46–52.
51. Guo Weihua and Yang Mingzeng. Monotonic stability of two-part series repairable system solutions. *Mathematics in Practice and Theory*, 2003, 33(4): 59–64.
52. Guo Weihua. Qualitative analysis to two identical parts cold standby repairable system solutions. *Acta Analysis Functionalis Applicata*, 2002, 4(4): 376–382.
53. Guo Weihua. Property analysis to two identical parts repairable man machine system solutions. *Mathematics in Practice and Theory*, 2003, 33(7): 88–96.
54. Wang Dingjiang. The existence of two identical parts parallel repairable system positive solutions. *Journal of Zhejiang University of Technology*, 2005, 33(3): 280–283.
55. Gao Dezhi. Posedness of a series dynamic maintenance system. *Mathematics in Practice and Theory*, 2003, 33(2): 80–85.
56. Guo Weihua. Monotonic stability of a single component repairable system solutions. *Journal of Qiongzhou University*, 2003, 10(5): 4–6.
57. Guo Weihua and Wu Songli. Solutions asymptotic stability of a repairable man-machine system with spare parts. *Mathematics in Practice and Theory*, 2004, 34(10): 104–110.
58. Guo Weihua. Existence and uniqueness of a two identical parts parallel repairable system solutions. *Mathematics in Practice and Theory*, 2002, 32(4): 632–634.
59. Wang Dingjiang. Stability of two identical parts parallel repairable system. *Journal of Zhejiang University of Technology*, 2006, 34(2): 228–229, 236.
60. Guo Weihua. Qualitative analysis to four-component redundant repairable system with common cause failure. *Journal of Zhoukou Normal University*, 2003, 20(2): 6–9.
61. Guo Weihua and Xu Genqi. Stability analysis to system consisting of robot and associated Safety device. *Mathematics in Practice and Theory*, 2003, 33(9): 116–122.
62. Zhang Guofen and Jiang Hongyan. Evaluation on non-parameter repairable system availability. *Journal of Zhejiang University*, 2005, 32(4): 377–381, 480.
63. Dyer D. Unification of reliability/availability/repairability models for Markov systems. *IEEE Transactions on Reliability*, 1989, 38(2): 246–252.
64. Rong Qunshan and Zhang Wenyu. Optimization model of preventive maintenance system. *Journal of Xi'an University of Post and Telecommunications*, 1998, 3(3): 69–73.
65. Reinek David M, Murdock Paul W and Pohl E A. Improving availability and cost performance for complex systems with preventive maintenance. *Proceedings Annual Reliability and Maintainability Symposium*, Washington, DC: IEEE, 1999, 383–388.
66. Barlow R E and Proschan F. *Mathematical theory of reliability*. New York: Wiley, 1965.
67. Wang Shaoping. *Engineering reliability*. Beijing: Beihang University Press, 2000.
68. Gao Shesheng and Zhang Lingxia. *Reliability theory and engineering applications*. Beijing: National Defend Industy Press, 2003.
69. Jin Xing and Hong Yanji. *System reliability and availability analysis method*. Beijing: National Defend Industy Press, 2007.
70. Zhao Jianmin. A Comprehensive optimization decision of preventive maintenance system. *Mechanical Science and Technology*, 2000, 19(4): 559–560.
71. Duarte J A C, Craveiro J C T A and Trigo T P. Optimization of the preventive maintenance plan of a series components system. *International Journal of Pressure Vessels and Piping*, 2006, 83(4): 244–248.

72. Limbourg Philipp and Hans-Dieter. K. Preventive maintenance scheduling by variable dimension evolutionary algorithms. *International Journal of Pressure Vessels and Piping*, 2006, 83: 262–269.

73. Jin Yulan, Jiang Zuhua and Hou Wenrui. Preventive maintenance strategies optimization of reliability-centered multi-component device. *Journal of Shanghai Jiaotong University*, 2006, 40(12): 2051–2056.

74. Das K, Lashkar R S and Sengupta S. Machine reliability and preventive maintenance planning for cellular manufacturing systems. *European Journal of Operational Research*, 2007, 183: 162–180.

75. Xie Hongzheng and Xie Hongwei. Approximation method and applications of nonlinear Volterra integral equation nontrivial solutions. *Acta Mathematicae Applicatae Sinica*, 1995, 18(4): 499–509.

76. Shen Yidan. *Integral equation*. Beijing: Beijing Institute of Technology Press, 1992.

77. Ball F B, Sansom M S. Single-channel auto-correlation functions: the effects of time interval omission. *Biophyical Journal*, 1988, 53(5): 819–832.

78. Hawkes A G, Jalali A and Colquhoun D. The distributions of the apparent open times and shut times in a single channel record when brief events can not be detected. *Philosophical Transactions: Physical Sciences and Engineering*, 1990, 332: 511–538.

79. Hawkes A G, Jalali A and Colquhoun D. Asymptotic distributions of apparent open times and shut times in a single channel record allowing for the omission of brief events. *Philosophical Transactions: Biological Sciences*, 1992, 337: 383–404.

80. Cui L R and Xie M. Availability analysis of periodically inspected systems with random walk model. *Journal of Applied Probability*, 2001, 38(4): 860–871.

81. Cui L R and Li J L. Availability for a repairable system with finite repairs. *Proceedings of the 2004 Asian International Workshop Advanced Reliability Modeling*. Hiroshima, Japan, World Scientific Publishing, 2004, 97–100.

82. Dagpunar J S. Renewal-type equations for a general repair process. *Quality and Reliability Engineering International*, 1997, 13(4): 235–245.

83. Chen Jingliang and Chen Xianghui. *Special matrix*. Beijing: Tsinghua University Press, 2001.

84. Zhong Yuquan. *Complex function theory*. Beijing: Higher Education Press, 2001.

85. Huang Lin. *Linear algebra in systems and control theory*. Beijing: Science Press, 1984.

86. Yang Yi. General probability distribution system instantaneous availability discrete time modeling analysis and application. PhD Thesis, Nanjing: Nanjing University of Science and Technology, 2008.

87. Lei Jigang, Tang Ping and Tian Ru. *Matrix theory and its applications*. Beijing: China MachinePress, 2005.

88. Horn R A, Johnson C R and Yang Qi. *Matrix analysis*. Beijing: China Machine Press, 2005.

89. Weibull W. A statistical distribution of wide applicability. *Applied Mechanics*, 1951, 18(3): 293–297.

90. Keshevan M K, Sargent G A and Conrad H. Statistical analysis of the Hertzian fracture of pyrex glass using the Weibull distribution function. *Materials Science*, 1980, 15(4): 839–844.

91. Sheikh A K, Boah J K and Hansen D A. Statistical modeling of pitting corrosion and pipeline reliability. *Corrosion*, 1990, 46(3): 3–8.

92. Queeshi F and Sheikh A K. A probabilistic characterization of adhesive wear in metals. *IEEE Transactions on Reliability*, 1997, 46(1): 38–44.
93. Durham S D and Padgett W J. A cumulative damage model for system failure with application to carbon fibers and composites. *Technometrics*, 1997, 39(1): 34–44.
94. Fork S L, Mitchell B C and Smart J, et al. A numerical study on the application of the Weibull theory to brittle materials. *Engineering Fracture Mechanics*, 2001, 68(10): 1171–1179.
95. Li S Q, Fang J Q and Liu D K, et al. Failure probability prediction of concrete components. *Cement and Concrete Research*, 2003, 33(10): 1631–1636.
96. Prabhakar Murthy D N, Bulmer Michael and Eccleston John A. Weibull model selection for reliability modeling. *Reliability Engineering and System Safety*, 2004, 86(3): 257–267.
97. Nadarajah S and Kotz S. On some recent modifications of Weibull distributions. *IEEE Transactions on Reliability*, 2005, 54(4): 561–562.
98. Bebbington M, Lai C D and Zitikis R. A flexible Weibull extension. *Reliability Engineering and System Safety*, 2007, 92(6): 719–726.
99. Pham H and Lai C D. On recent generalizations of the weibull distribution. *IEEE Transactions on Reliability*, 2007, 56(3): 454–458.
100. Wang Zhiming, Peng Anhua and Wang Qibing. Mechanical parts censored distribution reliability study. *Mei Kuang Jixie Coal Mine Machinery*, 2007, 28(1): 42–43.
101. Pan Ershun and Wang Shuyi. Parameter estimation based on truncated distribution in mechanical reliability design. *Machinery Design & Manufacture*, 1998, (3): 6–7.
102. Wang L C, Li J L and Zou Y. An improved annealing algorithm based on multi-agent. *Journal of Information and Computing Science*, 2006, 1(3): 161–167.
103. Eberhart R C and Kennedy J. A new optimizer using particles swarm theory. *Proceeding of 6th International Symposium on Micro Machine and Human Science*. Nagoya, Japan: Institute of Electrical and Electronics Engineers, 1995, 39–43.
104. Clieft M. The swarm and the queen: towards a deterministic and adaptive particle swarm optimization. *Proceedings of the IEEE Congress on Evolutionary Computation*. Washington, DC: IEEE, 1999, 1951–1957.

Index

279